RED GIANTS
AND WHITE DWARFS

"WE ARE BROTHERS OF THE BOULDERS,

IN AN UNBROKEN SEQUENCE OF EVENTS
THE UNIVERSE EXPANDS AND COOLS,
EARTH ARE FORMED, AND LIFE ARISES
ON THE SCENE.

A cloud of gas and dust in which stars are being born
(see also page 49).

COUSINS OF THE CLOUDS." Harlow Shapley

EXTENDING OVER TEN BILLION YEARS, STARS ARE BORN AND DIE, THE SUN AND ON THE EARTH. FINALLY, MAN APPEARS

RED GIANTS

MAN'S DESCENT

HARPER & ROW, PUBLISHERS

NEW YORK, EVANSTON, SAN FRANCISCO, LONDON

FROM THE STARS

REVISED EDITION

BY ROBERT JASTROW

DIRECTOR, GODDARD INSTITUTE FOR SPACE STUDIES

For information address Harper & Row, Publishers, Inc., 49 East 33rd Street, New York, N.Y. 10016. Published simultaneously in Canada by Fitzhenry & Whiteside Limited, Toronto.

STANDARD BOOK NUMBER: 06-012181-5

LIBRARY OF CONGRESS CATALOG CARD NUMBER: 79-108939

To my parents
with love

☼

Contents

Illustrations

☼

Preface to the Revised Edition

MAJOR scientific events in the last two years have required extensive revisions in the chapters of *Red Giants and White Dwarfs* dealing with astronomy and space science. The astronauts landed on the moon in 1969 and collected 80 pounds of lunar rock, which turned out, when analyzed in laboratories on the earth, to include the oldest materials ever found in the solar system. These findings confirmed what some scientists had long surmised: that the moon has preserved the record of its past far better than the earth, and holds clues to the early years of the solar system which are unavailable on our own planet. Mars and Venus were explored by unmanned Soviet and American spacecraft in several extraordinary flights between 1967 and 1969, which filled in blank areas in our knowledge of surface conditions on these planets and provided a better foundation for estimating the chances of finding life elsewhere in the solar system. Pulsars were discovered in 1968, and identified shortly thereafter as neutron stars—the densest objects known in the universe. The further discovery of a pulsar in the Crab Nebula tied pulsars to supernova explosions, and added an important and previously unsuspected element to the life story of the stars.

Completely new chapters have been written on the moon, Mars and Venus to include these fascinating results, and a new section on pulsars and neutron stars has been added to the chapter on stellar evolution. Other changes have been made throughout the text to reflect suggestions from readers and colleagues. It has been gratifying to receive a large and continuing volume of mail from readers around the world during the past three years, expressing their pleasure at being able to understand many of the mysteries of science for the first time through their reading of this book.

I am deeply indebted to Professor Paul W. Gast of the Lamont-Doherty Geological Observatory of Columbia University and the Manned Spacecraft Center in Houston for invaluable guidance in the interpretation of the findings from the Apollo 11 and Apollo 12 experiments and for many enjoyable discussions of the complexities of lunar geology. I should also like to thank Professor S. Ichtiaque Rasool of the Goddard Institute for Space Studies for several valuable discussions of the results of the Mariner missions to Mars.

Preface

This book had its genesis in a series of fifty-seven half-hour television programs that I did for the CBS network in 1964 as a morning lecture on space science called "Summer Semester." I decided to explain in these programs all fields of modern science, because the experiments carried on in the space program involve almost every important scientific problem, and I wanted the audience to understand why these experiments are worthwhile. I spanned most of the scientific spectrum—physics, astronomy, earth science and biology.

Toward the end of the series of programs I had a revelation that had never come to me before in fifteen years as a scientist. I realized that the latest advances in the separate fields of science, fascinating in themselves, are the brightly colored fragments of a mosaic which, when viewed from a distance, forms an image of the human observer himself, and of his origins. Science has uncovered evidence suggesting that we owe our existence to events which took place billions of years ago in stars that lived and died long before our solar system was formed. The scientific story of creation touches on the central problems of man's existence: What am I? How did I get here? What is my relation to the rest of the universe? The ideas are simple and beautiful; they can be expressed in clear language, without the use of jargon or mathematics. The story of man's origins goes far beyond the concepts of Darwin; it begins earlier than the time of our tree-dwelling ancestors, and much earlier than the period, several billion years ago, when the lowest forms of life first appeared on the face of the earth; it crosses the threshold between the living and the nonliving worlds and goes back in time to the parent cloud of hydrogen out of which all existing things are descended.

Caught in the grip of these fascinating ideas, I have put aside the rocket and satellite aspects of my earlier television programs, saving them for another volume, and have concentrated on a book that would tell of the evolution of stars, planets and life.

Many colleagues helped me in preparation of this volume. I am particularly indebted to several people who were kind enough to take the time required for a careful reading and the preparation of detailed criticisms of individual chapters. These include Professors Bengt Strömgren and A. G. W. Cameron and Dr. Richard Stothers for the chapters on the birth and death of stars and on cosmology; Professors Harold Urey, Gordon J. F. MacDonald and Paul W. Gast for the chapters on the origin of the solar system, the moon and the history of the earth; Dr. Gordon M. Tomkins and Professors Joshua Lederberg and Stanley L. Miller for the chapter on the origin of life; and Professors Edwin Colbert, John Imbrie, Armand V. Oppenheimer, Colin Pittendrigh and G. G. Simpson for the chapters on evolution.

I should like to express my special gratitude to a group of hard-working and devoted friends and associates. Chief among them are Kate Oliver and Ruth McCarthy, both blessed with clear minds and an excellent sense of language; the book owes a great deal to their many suggestions for the exposition of difficult scientific ideas. My mother, Marie Jastrow, read the manuscript carefully and pointed out remaining passages that might be difficult for the layman. Her suggestions appear throughout the book. I am grateful to Nicholas Panagakos for his excellent advice on problems of substance and exposition in every chapter. Nancy Stepan, Deborah Kaplan, Nancy Martin, Joseph Goldstein, Bonnie Neustadter and Alice Turner provided valuable advice regarding unclear passages. I should like to thank George Goodstadt and Barrett Gallagher for guidance in the selection and arrangement of illustrations. Barrett and his wife Timmy also made helpful suggestions which improved the manuscript, particularly in the earlier chapters on matter and forces. Philip Paris and Barbara Rusciolleli Stewart transcribed the audio tapes of the television programs into a first draft of the manuscript, and contributed many editorial improvements in that process. I am indebted to Inman King and Kay Roman for their assistance in the preparation of the manuscript.

Prologue

At the age of thirty I began to grow increasingly restless in my work. I had taught theoretical physics at Yale and Columbia, worked in the nuclear physics laboratory of the University of California in Berkeley, and spent some time at the Institute for Advanced Study in Princeton. Now I found myself still engaged in theoretical problems of nuclear structure, this time at the Naval Research Laboratory in Washington. My research had been limited to the fields of atomic and nuclear physics, and I had no feeling for the aspects of science that involved events outside the laboratory. Nonetheless I was intrigued by the activities of a group of scientists who were engaged in a completely different effort in another part of the Naval Research Laboratory. These were people working on Project Vanguard, the first American program for launching an artificial satellite.

The Vanguard project captured my imagination, and I decided to become involved in it if at all possible. With this end in mind I walked over one day in the fall of 1956 to offer my services to J. W. Siry, the chief theoretician of the project. Dr. Siry told me of a problem which he considered to be important, but which he could not take care of himself because of the burdens placed on him by the preparations for the first Vanguard launching. The problem he proposed was to calculate the slowing down of a satellite in orbit as a result of its collisions with the molecules of air. I did the calculations with a collaborator, Dr. C. A. Pearse. Our results suggested some peculiar electrical effects on the motion of the satellite. This research introduced me to the physics of satellite orbits.

In October, 1957, the Russians, and not the Americans, launched the first man-made satellite into space. Emboldened by the extraordinary event, I entered the Vanguard control room for the first time, to find the Vanguard radio networks tracking the Sputnik as it circled the earth. Accurate computing methods for the tracking of satellites were not yet ready, because the American program was geared to the launching of the Vanguard satellite scheduled to go up several months later.

The Vanguard team was under pressure to provide information on the position of the Sputnik. I asked whether I could help out with the preliminary calculations of the orbit. A naval ensign and a pretty math major recently out of Middlebury College were working at desks in a large room. Teletype machines clacked away with the results of radio positions of the Sputnik gathered from United States tracking stations stretched across North and South America. In the course of very rough calculations, I found the expected result that the Sputnik was steadily losing energy as a result of atmospheric friction. From the rate of energy loss the density of the earth's upper atmosphere could be calculated. Through this circumstance I first became acquainted with atmospheric physics.

The work on satellite orbits led to an invitation to participate in an International Rocket and Satellite Conference scheduled for July, 1958, in Moscow. At the same time I was given another assignment. Reports had been received of a bright object flying across the sky above Alaska and the western part of the United States, at around the time that the friction of the atmosphere would have brought back to earth the rocket which boosted the Sputnik into its orbit. The Russians announced that the rocket had come down over American territory, and that they wanted it back. Premier Khrushchev personally asked for it. Doubtless the United States government would have been delighted to accede to this request, but they did not have the rocket and were unable to find it.

I examined the available reports and radar sightings on the final orbits of the rocket and came to the conclusion that it had lasted in its orbit for six to eight hours after its final passage over the United States, and that most probably it had come down somewhere along a 2000-mile arc stretching between the eastern part of the Soviet Union and China.

On reaching the conference in Moscow in July, I presented my conclusions on the fate of the rocket. A pin could have been heard to drop in the hall as I delivered the dry summary of my calculations. No questions were asked by the several hundred Russian scientists who were present, and when the foreign correspondents filed their stories the Soviet censor held them up, although they were released the next day. To this day I do not know whether the Soviet scientists found the analysis convincing, or simply assumed that I had been ordered by the American government to manufacture my results.

From the Moscow conference I went on to the second United States Conference on the Peaceful Uses of Atomic Energy to discharge what turned out to be my last duty as a nuclear physicist. On returning to the United States in the fall, I was asked to join NASA, the newly created National Aeronautics and Space Administration, which had just been set up by the government in response to the Soviet challenge. My duty was to bring into existence a Theoretical Division of NASA, which was to devote itself to basic research in astronomy and the planetary sciences.

In a general way the area of research of the new division was obvious. It would include all scientific problems that could be studied by instruments carried on satellites and interplanetary rockets. But this definition of the job could have included most of science. The problem was to choose a few important problems and concentrate on them. What were the important problems?

I took my oath of appointment on November 10, 1958, and three weeks later I traveled across the United States to the laboratory at La Jolla, California, to visit a man who, I had been told, would be able to give me some advice. Professor Harold Clayton Urey had written a book on the moon and the planets, and was well known for the intensity of his interest in the scientific study of these objects. On arriving I introduced myself to him and asked him whether he could suggest some problems for the Theoretical Division of NASA to tackle, which would give us experience in a field so new to us.

Professor Urey seemed pleased to be sought out by a physicist working for the new space agency. He sat me down, handed me his book on the planets opened to the chapter on the moon, and began to tell me of the unique importance which this arid and lifeless body has for anyone who wishes to understand the origin of the earth and the other planets. I was fascinated by his story, which had never been told to me before in fourteen years of study and research in physics. Harold Urey has the marvelous quality of an intense, almost childlike, curiosity regarding all aspects of the natural world. This kind of curiosity is a rare quality. Through the eyes of Harold Urey, and others whom I met later, I first became acquainted with questions in science that were new to me, and completely different from the laboratory physics that had dominated my earlier training. As a graduate student, my life had revolved around the atom and the nucleus. In my world the laws of physics applied, and all events were governed by the action of the basic forces of nature—gravity, electromagnetism and the nuclear forces. Astronomy and the earth sciences—the study of the impact of these forces on a grand scale—were a closed book to me. Biology was an entirely alien subject.

Now I learned for the first time how stars and planets are born, how the

Harold Clayton Urey

solar system was formed, what conditions may have prevailed on the primitive earth, and how life may have developed on our planet. My previous work in physics had never led me to consider these matters. Yet the fact is that a single thread of evidence runs from the atom and the nucleus through the formation of stars and planets to the complexities of the living organism. Discoveries of the last decades link the world of the nucleus to the world of life in a chain of cause and effect which extends over billions of years, commencing with the formation of stars in our galaxy and ending with the appearance of man on the earth. At the beginning of this history there existed only atoms of the primeval element, hydrogen, which swirled through outer space in vast clouds. These clouds were the raw stuff out of which stars, planets and men were made. Occasionally the atoms of a cloud were drawn together by the attractive forces of gravity; with the passage of time the cloud contracted to a small, dense globe of gas; heated by self-compression, it rose in temperature until, at a level of some millions of degrees, its center burst into nuclear flame. Out of such events, stars were born.

Within the newborn star a series of nuclear reactions set in, in which all the other elements of the universe were manufactured out of the basic ingredient, hydrogen. Eventually these nuclear reactions died out, and the star's life came to an end. Deprived of its resources of nuclear energy, it collapsed under its own weight, and in the aftermath of the collapse an explosion occurred, spraying out to space all the materials that had been created within the star during its lifetime.

In the course of time, new stars, some with planets around them, condensed out of these materials. The sun and the earth were formed in this way, four and a half billion years ago, out of materials manufactured in the bodies of other stars earlier in the life of the Galaxy, and then dispersed to space when those stars exploded.

When the earth was first formed it must have been barren, but within one billion years or so life appeared on its surface. How can we explain this fact? What conditions prevailed on the earth during those first billion years? Recent discoveries hold clues to the answer. The biologists have discovered that certain molecules are the building blocks of all living creatures; they have created these molecules in the laboratory, out of the kinds of chemicals that existed in the atmosphere and oceans of the primitive earth; and they assert that life probably developed on the earth out of such molecules, several billion years ago.

Fossils preserved in the rocks of the earth's crust show that, during the next three billion years, a million varieties of plants and animals evolved out of these first living organisms. Why did those particular plants and animals arise, and not others? What forces pressed life into the forms it now possesses? Darwin's theory of evolution supplies an answer: Throughout the history of the earth the pressures of the struggle for existence have worked steadily on all creatures, shaping and molding the forms of life until each species acquires the best possible chance for survival in its environment. The fossil record displays the results of this process; it shows the gradual proliferation of many varieties of life, each specially adapted to one set of conditions. With the passage of time, and successive changes of climate, new forms of life spring up, and old ones are extinguished. At the end of the long chain of development man appears, as the product of a line of evolution that goes back far beyond his tree-dwelling ancestors, and even beyond the first forms of life on the earth. Man's history began billions of years before the solar system itself was formed; it began in a swirling cloud of primordial hydrogen. This is the history I shall relate in this book.

☼

1 The Size of Things

I ONCE had occasion to testify before the United States Senate Space and Aeronautics Committee on the scientific background of the space program. My talk dealt with the manner in which all substances in the universe are assembled out of neutrons, protons and electrons as the basic building blocks. After I left the chamber, a senior NASA official continued with a summary of the major space science achievements of the last year. Apparently my scholarly presentation had perplexed the senators, although they were anxious to understand the concepts I had presented. However, the NASA official's relaxed manner reassured them, and someone asked him: "How big is the electron? How much smaller is it than a speck of dust?" The NASA official correctly replied that the size of an electron is to a dust speck as the dust speck is to the entire earth.

The electron is indeed a tiny object. Its diameter is one 10-trillionth of an inch, a million times smaller than can be seen with the best electron microscope. Its weight is correspondingly small; 10,000 trillion trillion electrons make up one ounce. How can we be certain that such a small object exists? No one has ever picked up an electron with a pair of forceps and said, "Here is one." The evidence for its existence is all indirect. During the 150 years from the late eighteenth century to the beginning of the twentieth century a great variety of experiments were carried out on the flow of electricity through liquids and gases. The existence of the electron was not proved conclusively by any single one of these experiments. However, the majority of them could be explained most easily if the physicist assumed that the electricity was carried by a stream of small particles, each bearing its own electrical charge. Gradually physicists acquired a feeling, bordering on conviction, that the electron actually exists.

The question now was, how large is the electron, and how much electric charge does each electron carry? The clearest answer to this question came from an American physicist, Robert Millikan, who worked on the problem at the University of Chicago in the first decades of the twentieth century.

Millikan arranged a device, clever for its simplicity, in which an atomizer created a mist of very fine droplets of oil just above a small hole in the top of a container. A small number of the droplets fell through the hole and slowly drifted to the bottom of the container. Millikan could see the motions of these droplets very clearly by illuminating them from the side with a strong light so that they appeared as bright spots against a dark background. Millikan discovered that some of these droplets carried a few extra electrons, which had been picked up in the atomizing process. By applying an electrical force to the droplets and studying their motions in response to this force, he could deduce the amount of electric charge carried by the electrons on each droplet. This charge turned out to be exceedingly minute. As a demonstration of its minuteness, it takes an electric current equivalent to a flow of one million trillion electrons every second to light a 10-watt bulb. All this happened rather recently in the history of science. Millikan's first accurate measurements were completed in 1914.

The tiny *electron,* and two sister particles, are the building blocks out of which all matter in the world is constructed. The sister particles to the electron are the *proton* and the *neutron.* They were discovered even more recently than the electron; the proton was identified in 1920 and the neutron was first discovered in 1932. These two particles are massive in comparison with the electron—1840 times as heavy—but still inconceivably light by ordinary standards. The three particles combine in an amazingly simple way to form the objects we see and feel. A strong force of attraction binds neutrons and protons together to form a dense, compact body called the *nucleus,* whose size is somewhat less than one-trillionth of an inch. Electrons are attracted to the nucleus and circle around it as the planets circle around the sun, forming a solar system in miniature. Together the electrons and the nucleus make up the *atom.*

The size of a typical atom is one hundred-millionth of an inch. To get a feeling for the smallness of the atom compared to a macroscopic object, imagine that you can see the individual atoms in a kitchen table, and that each atom is the size of a grain of sand. On this scale of enlargement the table will be 2000 miles long.

The comparison of the atom with a grain of sand implies that the atom is a solid object. Actually, the atom consists largely of empty space. Each of the atoms that makes up the surface of a table consists of a number of electrons orbiting around a nucleus. The electrons form a diffuse shell around the nucleus, marking the outer boundary of the atom. The size of the atom is

10,000 times as great as the size of the nucleus at the center. If the outer shell of electrons in the atom were the size of the Astrodome that covers the Houston baseball stadium, the nucleus would be a Ping-pong ball in the center of the stadium. That is the emptiness of the atom.

If most of the atom is empty space, why does a tabletop offer resistance when you push it with your finger? The reason is that the surface of the table consists of a wall of electrons, the electrons belonging to the outermost layer of atoms in the tabletop; the surface of your finger also consists of a wall of electrons; where they meet, strong forces of electrical repulsion prevent the electrons in your fingertip from pushing past the outermost electrons in the top of the table into the empty space within each atom. An atomic projectile such as a proton, accelerated to high speed in a cyclotron, could easily pass through these electrons, which are, after all, rather light and unable to hurl back a fast-moving object. But it would take more force than the pressure of the finger can produce to force them aside and penetrate the inner space of the atom.

The concept of the empty atom is a recent development. Isaac Newton described atoms as "solid, massy, hard, impenetrable, moveable particles." Through the nineteenth century, physicists continued to regard them as small, solid objects. Lord Rutherford, the greatest experimental physicist of his time, once said, "I was brought up to look at the atom as a nice hard fellow, red or grey in color, according to taste." At the beginning of the twentieth century, J. J. Thomson, a British physicist and one of the pioneers in the investigation of the structure of matter, believed that the atom was a spherical plum pudding of positive electric charge in which negatively charged electrons were embedded like raisins. No one knew that the mass of the atom, and its positive charge, were concentrated in a small, dense nucleus at the center, and that the electrons circled around this nucleus at a considerable distance. But in 1911 Rutherford, acting on a hunch, instructed his assistant, Hans Geiger, and a graduate student named Marsden, to fire a beam of alpha particles into a bit of thin gold foil. These alpha particles are extremely fast-moving atomic projectiles which should have penetrated the gold foil and emerged from the other side. Most of them did, but Geiger and Marsden found that in a very few cases the alpha particles came out of the foil on the same side they had entered. Rutherford said later, "It was quite the most incredible event that has ever happened to me in my life. It was almost as incredible as if you fired a 15-inch shell at a piece of tissue paper and it came back and hit you." Later Geiger told the story of the occasion on which Rutherford saw the meaning of

Joseph J. Thompson (1856-1940)

the experiment. He relates: "One day (in 1911) Rutherford, obviously in the best of spirits, came into my room and told me that he now knew what the atom looked like and how to explain the large deflections of the alpha particles." What had occurred, Rutherford had decided, was that now and then an alpha particle hit a massive object in the foil, which bounced it straight back. He realized that the massive objects must be very small since the alpha particles hit them so rarely. He concluded that most of the mass of the atom is concentrated in a compact body at its center, which he named the nucleus. Rutherford's discovery opened the door to the nuclear era.

Let us continue with the description of the manner in which the universe is assembled out of its basic particles. Atoms are joined together in groups to form molecules, such as water, which consists of two atoms of hydrogen joined to one atom of oxygen. Large numbers of atoms or molecules cemented together form solid matter. There are a trillion trillion atoms in a cubic inch of an ordinary solid substance, which is roughly the same as the number of grains of sand in all the oceans of the earth.

The earth itself is an especially large collection of atoms bound together into a ball of rock and iron 8000 miles in diameter, weighing 6 billion trillion tons. It is one of nine planets, which are bound to the sun by the force of gravity. Together the sun and planets form the solar system. The largest of the planets is Jupiter, whose diameter is 86,000 miles; Mercury, the smallest, is 3100 miles across, one-third the size of the earth, and scarcely larger than the moon. All the planets are dwarfed by the sun, whose diameter is one million miles. The weight of the sun is 700 times greater than the combined weight of the nine planets. Like the atom, the solar system consists of a massive central body—the sun—surrounded by small, light objects—the planets—which revolve about it at great distances.

The sun is only one among 100 billion stars that are bound together by gravity into a large cluster of stars called the Galaxy. The stars of the Galaxy revolve about its center as the planets revolve about the sun. The sun itself participates in this rotating motion, completing one circuit around the Galaxy in 200 million years.

The Galaxy is flattened by its rotating motion into the shape of a disk, whose thickness is roughly one-fiftieth of its diameter. Most of the stars in the Galaxy are in this disk, although some are located outside it. A relatively small, spherical cluster of stars, called the nucleus of the Galaxy, bulges out of the disk at the center. The entire structure resembles a double sombrero with the galactic nucleus as the crown and the disk as the brim. The sun is located

A Photograph of Ernest Rutherford (1871-1937) Taken Outside the Cavendish Laboratory in 1934

in the brim of the sombrero about three-fifths of the way out from the center to the edge. When we look into the sky in the direction of the disk we see so many stars that they are not visible as separate points of light, but blend together into a luminous band stretching across the sky. This band is called the Milky Way.

The stars within the Galaxy are separated from one another by a distance of 30 trillion miles. In order to avoid the frequent repetition of such awkwardly large numbers, astronomical distances are usually expressed in units of the light year. A light year is defined as the distance covered in one year by a ray of light, which travels at 186,000 miles per second. This distance turns out to be six trillion miles; hence in these units the average distance between stars in the Galaxy is five light years, and the diameter of the Galaxy is 100,000 light years.

In spite of the enormous size of our galaxy, its boundaries do not mark the edge of the observable universe. The 200-inch telescope on Mount Palomar has within its range no less than 10 billion other galaxies, each comparable to our own in size and containing a similar number of stars. The average distance between these galaxies is one million light years. The extent of the visible universe, as it can be seen in the 200-inch telescope, is 10 billion light years.

An analogy will help to clarify the meaning of these enormous distances. Let the sun be the size of an orange; on that scale of sizes the earth is a grain of sand circling in orbit around the sun at a distance of 30 feet; the giant planet Jupiter, 11 times larger than the earth, is a cherry pit revolving at a distance of 200 feet or one city block; Saturn is another cherry pit two blocks from the sun; and Pluto, the outermost planet, is still another sand grain at a distance of ten city blocks from the sun.

On the same scale the average distance between the stars is 2000 miles. The sun's nearest neighbor, a star called Alpha Centauri, is 1300 miles away. In the space between the sun and its neighbors there is nothing but a thin distribution of hydrogen atoms, forming a vacuum far better than any ever achieved on earth. The Galaxy, on this scale, is a cluster of oranges separated by an average distance of 2000 miles, the entire cluster being 20 million miles in diameter.

An orange, a few grains of sand some feet away, and then some cherry pits circling slowly around the orange at a distance of a city block. Two thousand miles away is another orange, perhaps with a few specks of planetary matter circling around it. That is the void of space.

THE NUCLEUS AND THE ATOM. Matter is composed of three fundamental building blocks: a light particle called the electron and two relatively heavy particles, each 1840 times as massive as the electron, called the neutron and the proton. The neutrons and protons are bound very tightly into a compact mass called the nucleus.

Electrons are attracted to the nucleus by the positive charge of the protons and circle in orbits around it under the influence of this attraction. Together the nucleus and orbiting electrons form the atom. An atom of carbon *(above)* consists of six electrons circling a nucleus of six protons and six neutrons. The diameter of a typical atom is one hundred-millionth of an inch.

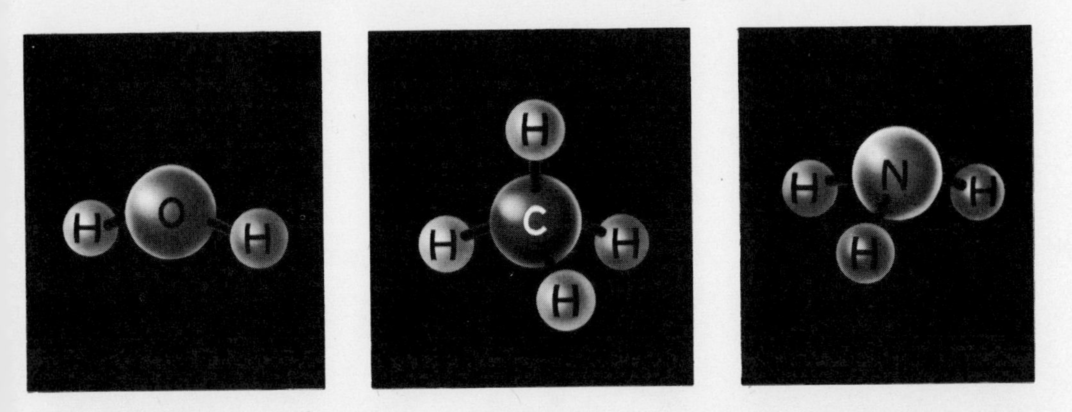

WATER METHANE AMMONIA

MOLECULES AND SOLID MATTER. Atoms are bound together in groups to form molecules. A molecule of water consists of two atoms of hydrogen bound to one atom of oxygen. Other compounds of hydrogen with common elements are methane (a carbon atom bound to four hydrogen atoms) and ammonia (a nitrogen atom bound to three hydrogen atoms).

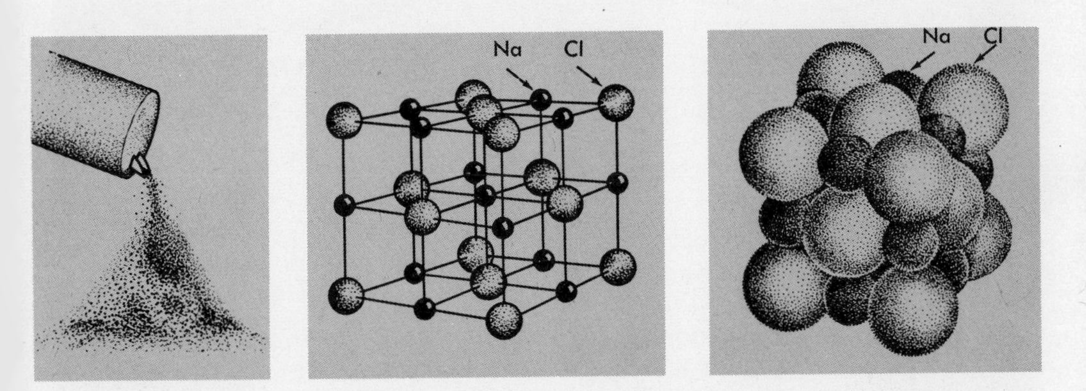

Large numbers of atoms or molecules joined together form solid matter. Crystals of common salt *(left)* consist of atoms of sodium (Na) and chloride (Cl) arranged in a lattice *(middle)*. One grain of salt contains a million trillion atoms. In the drawing of the lattice, large spaces are shown between the atoms for clarity. In actuality, they are closely packed *(right)*.

THE EARTH. The earth is a large collection of atoms cemented together in a ball of rock 8000 miles in diameter and weighing 6 billion trillion tons. At the center is a core of molten nickel and iron 1800 miles in radius. Surrounding the molten core is a mantle of solid rock 2200 miles thick. The mantle is capped by a crust of lighter rocks whose average thickness is about 15 miles.

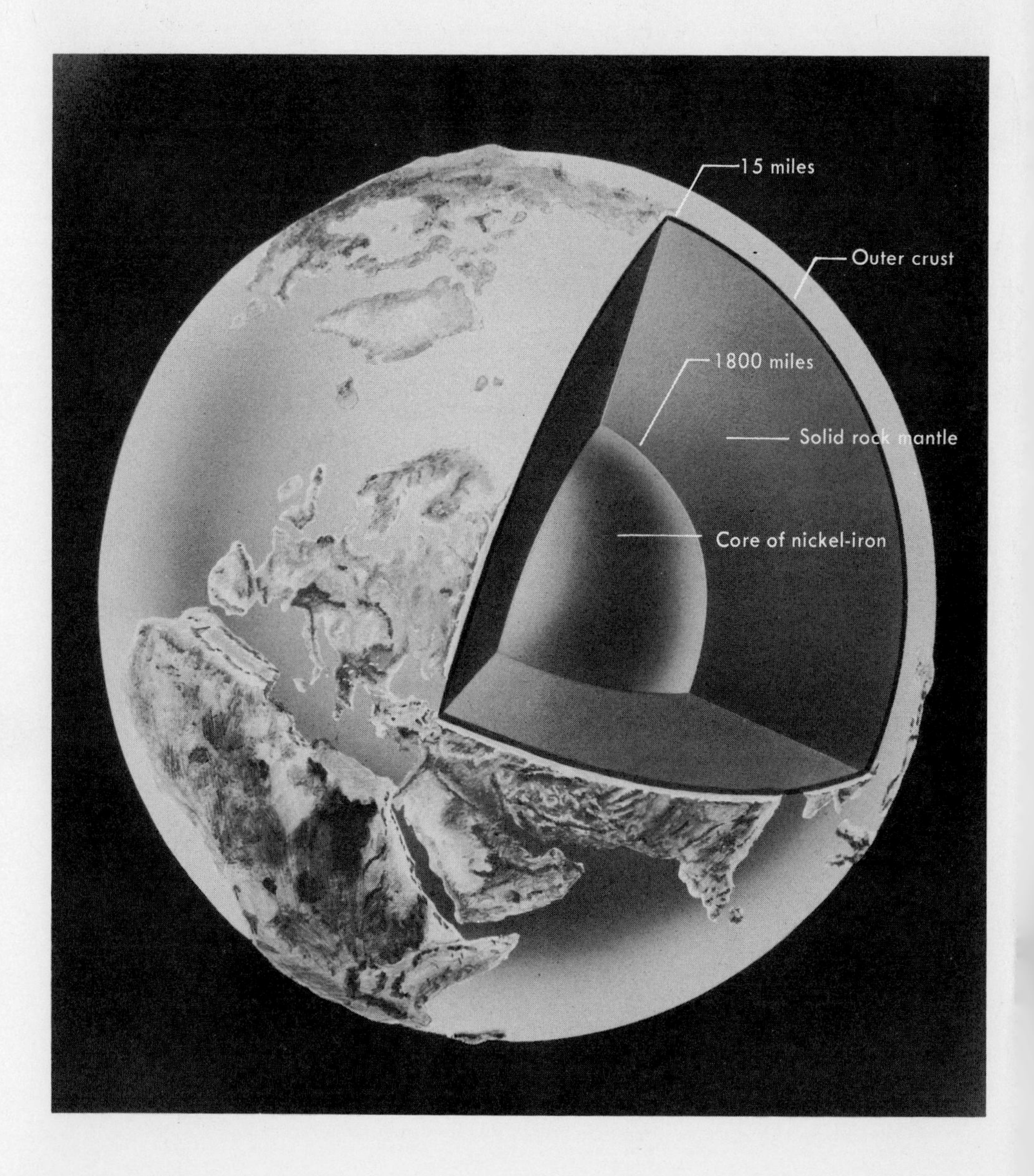

THE SOLAR SYSTEM. The earth is one of nine planets that revolve in orbits around the sun, bound to it by the force of gravity. The weight of the sun is 700 times greater than the combined weight of the nine planets. The sun, the planets, their thirty-two satellites, and a large number of other lesser bodies, including asteroids and comets, form the solar system.

The orbits of all the planets are within a few degrees of one plane, except for the orbits of the innermost planet, Mercury, and the outermost planet, Pluto. All orbits are close to perfect circles, with the exception, again, of Mercury and Pluto. Pluto circles the sun at a distance varying from 3 billion to 4 billion miles. The orbit of Pluto marks the outer boundary of the solar system.

In the space between the planets, and outside the boundary of the solar system in the space between the stars, there is a tenuous cloud of hydrogen gas with a density of 10 to 100 atoms per cubic inch. The distance from the sun and its planets to the nearest known star, Alpha Centauri, is 25 trillion miles, or 5000 times the size of the solar system. Outer space, like the inner space of the atom, is nearly empty.

THE SOLAR SYSTEM

PLUTO
MERCURY
VENUS
SUN
MARS
MOON
EARTH
ASTEROID BELT
SATURN

OUR GALAXY. The sun is one of 100 billion stars bound together by the force of gravity into a large cluster called the Galaxy. Only a few thousand of these stars can be seen with the naked eye. However, many more appear in photographs taken with a large telescope. Approximately 10,000 stars are visible in the photograph at right, although it represents only a thousandth of the area of the full night sky. This photograph was obtained with a 10-inch telescope by an exposure of several hours.

The Galaxy is flattened by its spinning motion into the shape of a disk, with the sun located roughly halfway from the center of the disk to the edge. The artist's diagram *(below)* shows the structure of the Galaxy, viewed edge-on, with the position of the sun indicated.

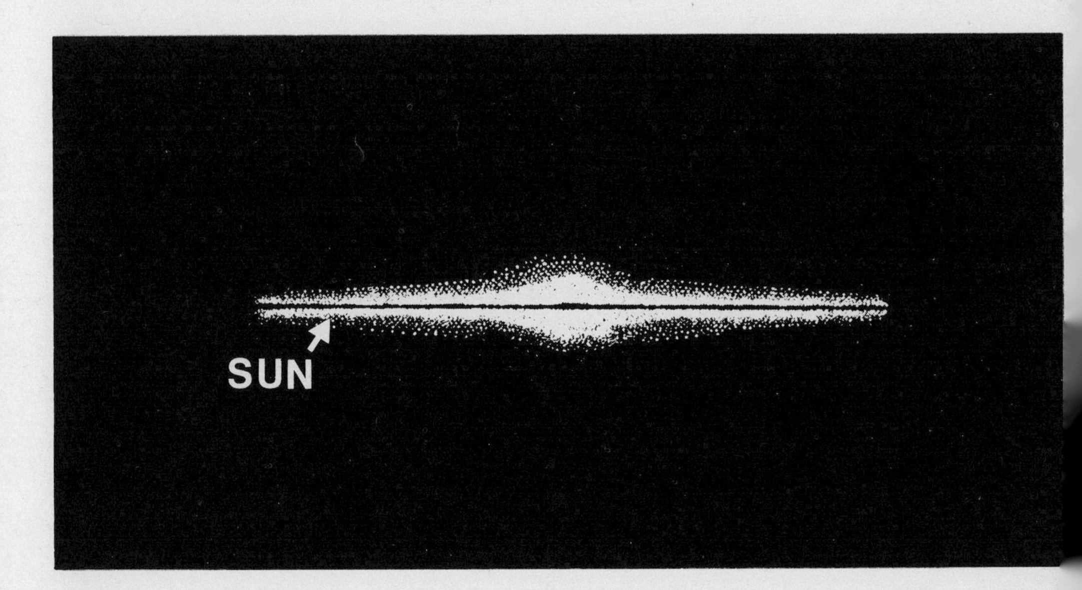

The distances between stars are usually expressed in terms of the light year, which is the distance covered in one year by a ray of light traveling at 186,000 miles per second. One light year is 6 trillion miles. The average distance between stars in our part of the Galaxy is 6 light years. The diameter of the entire Galaxy is 100,000 light years, and the thickness of the central disk is about 2000 light years.

The concentration of stars in the central disk of the Galaxy makes it an exceedingly bright region when viewed edge-on. On a clear evening, if we are away from city lights, we see this band of light stretching across the sky. It is called the Milky Way.

A montage of photographs displaying the Milky Way appears on the following pages.

STARS IN OUR GALAXY

THE MILKY WAY: AN EDGE-ON VIEW OF OUR GALAXY. This montage of photographs shows the Galaxy viewed edge-on, as it would appear to an observer in our solar system. The 7000 brightest stars have been drawn in separately. The bright, luminous clouds con-

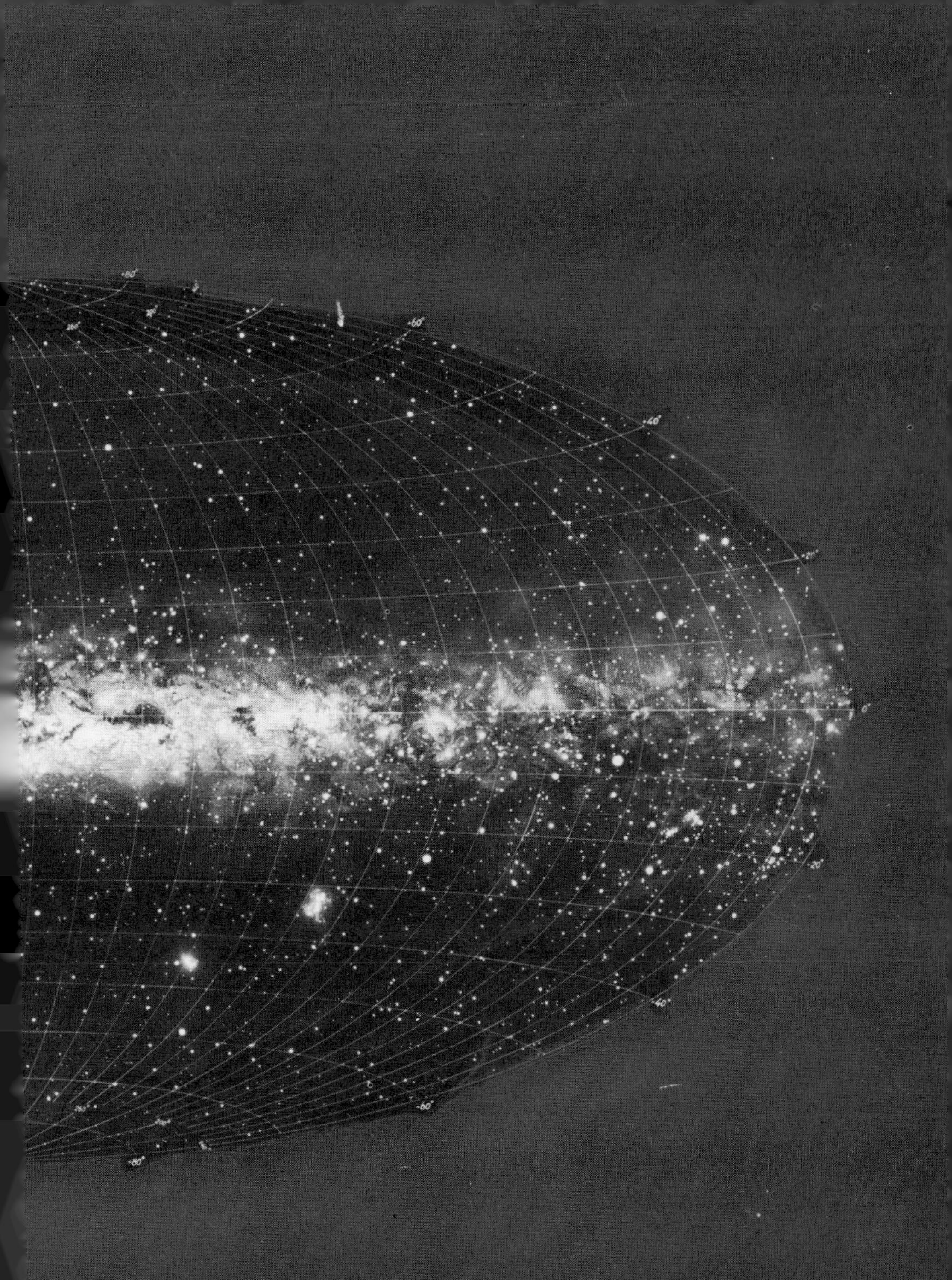

sist of billions of individual stars. The dark regions running irregularly through the center of the Milky Way are clouds of dust which absorb the light coming to our solar system from distant stars in the Galaxy.

NEIGHBORING GALAXIES. Our nearest extragalactic neighbors are two small galaxies, satellites of our own, which are held captive by the gravitational force of the 100 billion stars in our galaxy. Each of these satellite galaxies contains a few billion stars. They are visible to the naked eye as faint patches of light, called the Magellanic Clouds.

Roughly a dozen other galaxies exist within 3 million light years of our galaxy. Their positions are shown below, together with the titles assigned to them in astronomical catalogues.

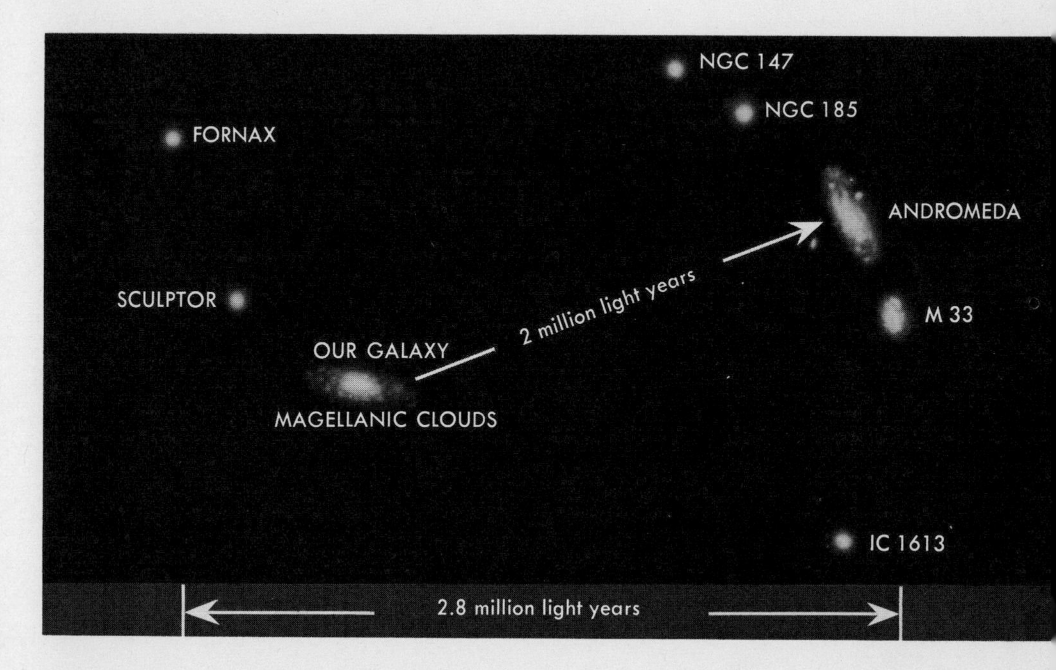

The nearest galaxy of comparable size to our own is the Great Nebula in Andromeda, located approximately 2 million light years from us. The Andromeda galaxy closely resembles ours in size, shape and number of stars. At the right is a photograph of the Andromeda galaxy taken with the 200-inch telescope on Mount Palomar. The small points of light are individual stars within our own galaxy, lying in the line of sight to Andromeda. The two luminous regions above and below Andromeda are satellite galaxies, similar to the Magellanic Clouds.

The Andromeda Galaxy

GALACTIC SHAPES. The shape of our galaxy is illustrated clearly in the sequence opposite, which shows four different disk-shaped galaxies, all resembling ours but tilted at various angles to our line of sight. The galaxies are identified by the catalogue numbers assigned to them by astronomers. The sun is located in a galaxy similar to these at a point approximately halfway from the center to the edge.

The fourth photograph shows the spiral arms characteristic of the type of galaxy to which the sun belongs. These spiral arms are concentrations of gas and dust in which stars are born.

Not all galaxies have this shape. Some are round, some are elliptical, and some are irregular. A peculiar galaxy, M82, is shown below. M82 is 10 million light years from our galaxy. A study of the radiation emitted from this galaxy indicates that jets of hot hydrogen gas are shooting out from its center. M82 is one of several "exploding galaxies"; there is evidence that similar explosions are taking place at the center of our own galaxy. Their cause is not known.

NGC 4565 Viewed edge-on

NGC 4216 Tilted 15 degrees

NGC 7331 Tilted 30 degrees

NGC 628 Viewed face-on

CLUSTERS OF GALAXIES. Ten billion other galaxies are within range of the 200-inch telescope on Mount Palomar. Most of these galaxies occur in clusters, ranging from as few as 3 to as many as 10,000 galaxies in one cluster. The galaxies in a cluster are separated by an average distance of 1 million light years. Our galaxy is a member of a small cluster of galaxies called the Local Group, which includes the large galaxy in Andromeda and approximately a dozen others. Several galaxies in the Local Group are shown in the drawing on page 24.

The photograph opposite shows a group of approximately 50 galaxies at the center of the giant cluster of galaxies in the Constellation Hercules, about 300 million light years distant. The Hercules cluster, containing upward of 10,000 galaxies, is one of the major systems of organized matter in the universe. In the photograph, the spiked objects and some of the perfectly circular spots are individual stars in our galaxy; every other object is a galaxy in the Hercules cluster.

Clusters of galaxies are the largest known systems of organized matter. They terminate the hierarchy of structure in the universe.

A CLUSTER OF GALAXIES

☼

2 The Forces of Nature

> There are therefore Agents in Nature able to make the Particles of Bodies stick together by very strong Attractions. And it is the Business of experimental Philosophy to find them out.
>
> —ISAAC NEWTON, *Opticks* (1704)

IT IS remarkable that all objects in the universe, from the smallest nucleus to the largest galaxy, are held together by only three fundamental forces—a nuclear force, the force of electromagnetism, and the force of gravity.

Most powerful is the nuclear force,* which binds neutrons and protons together into the nucleus of the atom. This extremely strong force of attraction pulls the particles of the nucleus together into an exceedingly compact body with a density of one billion tons per cubic inch.

Next strongest is the electrical (electromagnetic) force, which is approximately 100 times weaker than the nuclear force. This force binds the electrons to the nucleus to form atoms, and binds atoms together into solid matter.

Least powerful is the force of gravity. The gravitational force is exceedingly weak, about 10^{38} times weaker than the nuclear force, and 10^{36} times weaker than the force of electricity.† These numbers are exceedingly large; 10^{36} is a trillion times larger than the number of grains of sand in the oceans of the earth. Nonetheless it is this very frail agent which keeps the moon in orbit around the earth, the earth and other planets revolving around the sun, and the sun and other stars clustered together in our galaxy.

In one sense gravity is the simplest of the basic forces. Its simplicity lies in the fact that the force of gravity acting between two objects always pulls them

* Two kinds of nuclear force exist—the one discussed here, and a weaker one, called the weak force. The latter plays no substantial role in this discussion, and I omit it for simplicity.

† 10^{38} is 1 followed by 38 zeros; one says: "10 to the 38th." It is a convenient shorthand for writing very large numbers. Some of these numbers have names. For example, 10^3 (= 1000) is one thousand; $10^6 = 1$ million; $10^9 = 1$ billion; $10^{12} = 1$ trillion = 1 million million. Numbers larger than this, however, do not have names in common usage.

together and never pushes them apart. The electrical force is more complicated because its effect on some pairs of particles is to pull them together, while on others its effect is to push them apart. The explanation of these two kinds of action has come out of laboratory experiments with electricity during the last 200 years. These experiments show that two types of electricity exist, called positive and negative. When two particles bear the same kind of charge, i.e., both negative or both positive, they repel each other; but two particles bearing different kinds of charge are attracted to one another.

Every electron carries a charge of negative electricity; therefore all electrons repel one another. Similarly, every proton carries a charge of positive electricity; therefore all protons repel one another. However, the electron and the proton attract one another because they carry different kinds of electricity.

This last fact enables us to understand the existence of the atom. The nuclear force acts as a glue which binds the neutrons and protons together to form the nucleus. The protons within the nucleus, being positively charged, exert an electrical attraction on any electrons that may be in the neighborhood. Under the influence of this attraction electrons are captured by the nucleus and forced to circle around it, as the planets circle the sun under the attraction of gravity. The nucleus and its surrounding electrons together constitute the atom.

The simplest atom consists of a single electron circling around a nucleus composed of a single proton. This is the atom of hydrogen, the most abundant element found in nature, which makes up 90 percent of all matter in the universe.

The neutron is the other basic nuclear particle; it carries no electric charge, and is not affected by electrical forces. However, neutrons are attracted to protons by the very strong nuclear force. Under the influence of this attraction, a neutron and a proton may be joined into a single, heavier nucleus, weighing twice as much as the proton. The new nucleus contains one positive charge, and therefore attracts a single electron to itself, forming an atom similar to hydrogen, but twice as massive. This atom is called heavy hydrogen, or deuterium. Heavy hydrogen was discovered by Harold Urey in 1932, an achievement for which he received the Nobel Prize in 1934. It enters into combination with oxygen to form heavy water, a substance identical with water but somewhat denser. Heavy water is a relatively scarce substance, there being one molecule of heavy water to each 10,000 molecules of ordinary water in the oceans.

After hydrogen, the second most abundant element in the universe is helium. Helium usually exists on the earth in the form of a gas. It is lighter than air, once had a wide use in dirigibles, and is still used in children's balloons. The nucleus of a helium atom contains two neutrons and two protons. Since each proton can attract one electron to itself, an atom of helium has two electrons circling in orbit around the nucleus.

Hydrogen and helium together constitute approximately 99 percent of the matter in the universe. All other elements make up the remaining one percent. Among the elements in this last one percent, the most abundant is the critically important substance, oxygen. An atom of oxygen is composed of a nucleus containing eight neutrons and eight protons, around which eight electrons circle. Other nuclei exist with still greater numbers of neutrons and protons. In each case, the complete atom is formed when the nucleus has circling around it a number of electrons equal to the number of protons it contains. In the heaviest elements, such as lead and gold, the nucleus contains approximately 200 neutrons and protons, surrounded by up to 92 electrons arranged in a complicated array of orbits of many different sizes. The atom of uranium is the largest, heaviest and most complicated of all. It consists of a nucleus containing 146 neutrons and 92 protons, surrounded by 92 electrons.

Altogether there exist 92 different kinds of nuclei and 92 corresponding kinds of atoms, stretching from hydrogen to uranium. These elements, in various combinations, make up all the varieties of matter—animal, vegetable and mineral—which exist in the universe.

Hydrogen atom

Heavy hydrogen atom (deuterium)

Helium atom

Carbon atom

SIMPLE ATOMS. The diagram shows atoms of hydrogen, heavy hydrogen, helium and carbon. Each atom consists of a nucleus composed of neutrons and protons, surrounded by electrons which orbit the nucleus at a considerable distance. The protons carry a positive electric charge which attracts negatively charged electrons until the number of electrons orbiting around the nucleus is equal to the number of protons it contains. Each hydrogen atom contains one electron, which orbits the single proton in the nucleus; the helium atom contains two orbiting electrons and two nuclear protons; the carbon atom contains two electrons in an inner orbit and four electrons in an outer orbit, or six in all, balancing the six protons in the carbon nucleus.

Most properties of an element depend on the number of electrons in the atom; this number is in turn fixed by the number of protons in the nucleus. Thus the nucleus ultimately determines the properties of all elements.

☼

3 The Alchemist's Goal

THE ancients believed that all the varieties of matter in the world could be made by mixing together the four basic substances—earth, air, fire and water—in the right proportions and with suitable amounts of dryness, moisture, heat and cold. A strong element of mysticism often pervaded the experiments. Correct proportions were not enough; suitable incantations and recipes had to be followed. These beliefs nourished the science of alchemy throughout the Middle Ages down to the time of Isaac Newton.

Newton himself was an enthusiastic alchemist. His nephew and amanuensis, Humphrey Newton, wrote, "About six weeks in the spring and six weeks in the fall, the fire in the laboratory scarcely went out. . . . he would sometimes look into an old moldy book which lay in his laboratory. I think it was called *Agricola de Metallis,* the transmuting of metals being his chief design. . . ." Newton appears to have taken greater pleasure in these alchemical experiments than in the labors in mathematics and physics which secured his fame. He believed in the transmutation of elements, specifically of lead into gold, and was interested in the money-making possibilities of such endeavors. In 1669 he wrote in a letter to his friend Francis Aston, "And if you meet with any transmutations out of their own species into another . . . those, above all, will be worth your noting, being the most *luciferous,* and many times *lucriferous* experiments in philosophy." "Lucriferous" was a term in usage among alchemists to describe the potentially profitable applications of their science, as opposed to the luciferous applications—those which added to the sum of knowledge.

Newton's exceptional talents led him to the invention of the calculus and the discovery of the law of universal gravity. Yet the architect of the universe failed completely in his effort to transmute the elements. The reasons for his failure could not have been perceived by anyone in the seventeenth century. They did not become apparent until Rutherford's discoveries in the first years of the twentieth century. Then it became clear that Newton and the alchemists had been tampering only with the superficial qualities of matter. They were not striking at the heart of the problem—the nucleus.

It is in the nucleus that the quintessence of a substance resides. What is the difference between lead and gold? Each is a heavy metal; one is base, common and of a dull gray color, the other noble, rare and a lustrous yellow. Yet the two elements are surprisingly close in the hierarchy of atomic structure. An atom of gold is composed of a nucleus containing 118 neutrons and 79 protons, around which 79 electrons circle in orbit. An atom of lead contains 126 neutrons and 82 protons surrounded by 82 electrons. The eight additional neutrons in the lead nucleus serve only to make it a somewhat heavier element than gold; but the three additional protons play a more important role, for they draw three extra electrons to the lead atom. It is these extra electrons which are felt, seen and touched; they are responsible for the difference between lead and gold. Yet the number of electrons in an atom is controlled entirely by the number of protons in the nucleus. If three of the protons in the nucleus of an atom of lead could be removed in some way, the resulting nucleus, with 79 protons, would be the nucleus of an atom of gold.

But it is not possible to dislodge protons from the nucleus very easily. The fires of the alchemist's furnace can do no more than melt a bar of lead by breaking the bonds between neighboring atoms. Perhaps, if the flame is hot enough, some of the atoms may lose an electron. But an atom stripped of one electron is still an atom of lead. It still contains 82 protons, and if it lacks an electron it will exert an electrical attraction on other electrons nearby until it has collected the full complement of 82 to become a normal atom of lead again.

The nature of an element can be changed only by adding protons to the nucleus or removing them. But the force that keeps the proton within the nucleus is far stronger than the force that binds the electron to the atom or the force that cements atoms together into solid matter. Because the proton is locked so firmly within the nucleus, an exceptional degree of force must be applied to dislodge it. An ordinary blow with a hammer will not do, because the hammer cannot smash through the electrons of the atom to reach the nucleus at the center. Only an encounter with another minute particle, moving at speeds great enough to penetrate the electron barrier around the nucleus, will suffice.

Such an encounter was observed for the first time by Rutherford in a classic experiment carried out in 1919, in which he bombarded nitrogen atoms with high-speed helium nuclei emitted from radium. In a few cases, a fast-moving helium nucleus collided directly with a nitrogen nucleus and stuck to it to form a heavier particle, which was the nucleus of a new element.

What new element was created by Rutherford? The nitrogen nucleus contains seven neutrons and seven protons; the helium nucleus contains two neutrons and two protons. The nucleus resulting from their union therefore contains nine neutrons and nine protons. Rutherford found that one of the protons was dislodged by the violence of the impact and left the scene of the collision, leaving a residual nucleus with eight protons. The nucleus then collected eight electrons in orbit around it to become an atom of *oxygen*. This atom, containing nine neutrons, was slightly heavier than the common variety of oxygen, whose nucleus contains only eight neutrons, but it was, nonetheless, a bona-fide oxygen atom. The experiment had transmuted nitrogen into a different substance, oxygen.

Of course the yield of the newly created element was exceedingly small. Only one helium nucleus in 50,000 collided with a nitrogen nucleus and fused with it to create oxygen; hence, very few oxygen atoms resulted from the experiment. Nevertheless, Rutherford's experiment was a breakthrough, for it showed the way in which new elements can be created by building up large nuclei out of smaller ones. We suspect today that all the elements in the sun and the planets, and in our bodies, were created in the centers of other stars, earlier in the history of the Galaxy, by the same kind of nuclear bombardment which Rutherford performed in the laboratory.

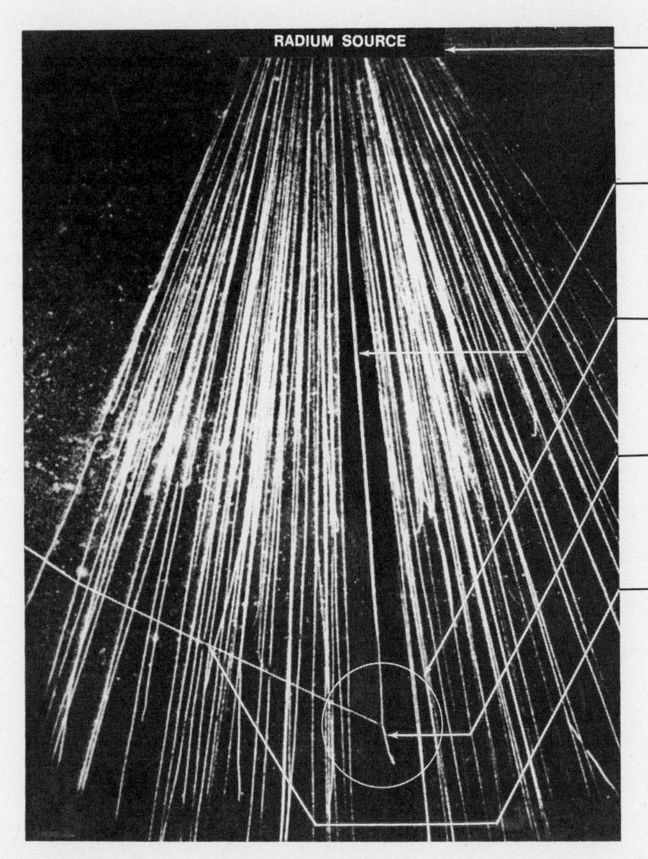

1. Tracks of helium nuclei emerge from radium source at top.

2. Arrow points to track of helium nucleus which will undergo collision below.

3. Circle marks site of collision between helium nucleus and nitrogen nucleus in cloud chamber.

4. Short track moving downward to right is produced by nucleus recoiling from collision.

5. Proton knocked out during collision moves upward to left

(Some helium tracks have been removed to exhibit the track of the nucleus involved in the collision more clearly.)

RUTHERFORD'S TRANSMUTATION OF AN ELEMENT. In order to change the basic properties of an element, it is necessary to change its nucleus. The photograph above was obtained in one of the first experiments performed by Rutherford in his Cambridge laboratory in which one element was artificially transformed into another. The photograph shows a cloud chamber—a vessel filled with water vapor and illuminated from the side to show the tracks of particles passing through it. A thin foil of radium, shown schematically at the top, emits fast-moving helium nuclei at a steady rate as the products of a radioactive decay process. Some fifty tracks produced by these helium nuclei are visible, spreading downward through the chamber from the radium foil (1).

The cloud chamber contains nitrogen gas in addition to water vapor. Among the many helium nuclei which pass through the chamber, one (2) collides with the nucleus of an atom of nitrogen and fuses with it to form a new nucleus. The collision occurs at the center of the circle (3). The resultant nucleus is that of oxygen; its track may be seen as a short stub moving downward and slightly to the right within the circle (4). The force of the impact has dislodged a proton, whose track is visible moving to the left and upward (5).

The artificial transmutation of nitrogen into oxygen by Rutherford, as demonstrated in this cloud-chamber photograph, marked the first time that man achieved the alchemist's goal by changing one element into another.

☼

4 Red Giants, White Dwarfs and Pulsars

THE stars seem immutable, but they are not. They are born, evolve and die like living organisms. The life story of a star begins with the simplest and most abundant element in nature, which is hydrogen. The universe is filled with thin clouds of hydrogen, which surge and eddy in the space between the stars. In the swirling motions of these tenuous clouds, atoms sometimes come together to form small pockets of gas. These pockets are temporary condensations in an otherwise highly rarefied medium. Normally the atoms fly apart again in a short time as a consequence of their random motions, and the pocket of gas quickly disperses to space. However, each atom exerts a small gravitational attraction on its neighbor, which counters the tendency of the atoms to fly apart. If the number of atoms in the pocket of gas is large enough, the accumulation of all these separate forces will hold it together indefinitely. It is then an independent cloud of gas, preserved by the attraction of each atom in the cloud to its neighbor.

With the passage of time, the continuing influence of gravity, pulling all the atoms closer together, causes the cloud to contract. The individual atoms "fall" toward the center of the cloud under the force of gravity; as they fall, they pick up speed and their energy increases. The increase in energy heats the gas and raises its temperature. The shrinking, continuously self-heating ball of gas is an embryonic star.

As the gas cloud contracts under the pressure of its own weight, the temperature at the center mounts steadily. When it reaches 100,000 degrees Fahrenheit, the hydrogen atoms in the gas collide with sufficient violence to dislodge all electrons from their orbits around the protons. The original gas of hydrogen atoms, each consisting of an electron circling around a proton, becomes a mixture of two gases, one composed of electrons and the other of protons.

At this stage the globe of gas has contracted from its original size, which was 10 trillion miles in diameter, to a diameter of 100 million miles. To understand the extent of the contraction, imagine the Hindenburg dirigible shrinking to the size of a grain of sand.

The huge ball of gas—now composed of separate protons and electrons—continues to contract under the force of its own weight, and the temperature at the center rises further. After 10 million years the temperature has risen to the critical value of 20 million degrees Fahrenheit.* At this time, the diameter of the ball has shrunk to one million miles, which is the size of our sun and other typical stars.

Why is 20 million degrees a critical temperature? The explanation is connected with the forces between the protons in the contracting cloud. When two protons are separated by large distances, they repel one another electrically because each proton carries a positive electric charge. But if the protons approach within a very close distance of each other, the electrical repulsion gives way to the even stronger force of nuclear attraction. The protons must be closer together than one 10-trillionth of an inch for the nuclear force to be effective. Under ordinary circumstances, the electrical repulsion serves as a barrier to prevent as close an approach as this. In a collision of exceptional violence, however, the protons may pierce the electrical barrier which separates them, and come within the range of their nuclear attraction. Collisions of the required degree of violence first begin to occur when the temperature of the gas reaches 20 million degrees.

Once the barrier between two protons is pierced in a collision, they pick up speed as a result of their nuclear attraction and rush rapidly toward each other. In the final moment of the collision the force of nuclear attraction is so strong that it fuses the protons together into a single nucleus. At the same time the energy of their collision is released in the form of heat and light. This release of energy marks the birth of the star.

The energy passes to the surface and is radiated away in the form of light, by which we see the star in the sky. The energy release, which is one million times greater per pound than that produced in a TNT explosion, halts the further contraction of the star, which lives out the rest of its life in a balance between the outward pressures generated by the release of nuclear energy at its center and the inward pressures created by the force of gravity.

The fusion of two protons into a single nucleus is only the first step in a

* Twenty million degrees is a very high temperature. For comparison, the temperature of the flame in the gas burner of the kitchen stove is 1,000 degrees, and the temperature of the hottest steel furnace is 10,000 degrees.

series of reactions by which nuclear energy is released during the life of the star. In subsequent collisions, two additional protons are joined to the first two to form a nucleus containing four particles. Two of the protons shed their positive charges to become neutrons in the course of the process. The result is a nucleus with two protons and two neutrons. This is the nucleus of the helium atom. Thus, the sequence of reactions transforms protons, or hydrogen nuclei, into helium.*

The fusion of hydrogen into helium is the first and longest stage in the history of a star, occupying about 99 percent of its lifetime. Throughout this long period of the star's life its appearance changes very little, but toward the end of the hydrogen-burning stage, when most of the hydrogen has been converted into helium, the star begins to show the first signs of age. The telltale symptoms are a swelling and reddening of the outer layers, commencing imperceptibly and progressing until the star has grown to a huge red ball 100 times larger than its original size. The sun will reach this stage in another 5 billion years, at which time it will have swollen into a vast sphere of gas engulfing the planets Mercury and Venus and reaching out nearly to the orbit of the earth. This red globe will cover most of the sky when viewed from our planet. Unfortunately we will not be able to linger and observe the magnificent sight, because the rays of the swollen sun will heat the surface of the earth to 4000 degrees Fahrenheit and eventually evaporate its substance. Perhaps Jupiter will be a suitable habitat for us by then. More likely, we will have fled to another part of the Galaxy.

Such distended, reddish stars are called *red giants* by the astronomers. An example of a red giant is Betelgeuse, a fairly bright star in the constellation Orion, which appears distinctly red to the naked eye.

A star continues to live as a red giant until its reserves of hydrogen fuel are exhausted. With its fuel gone, the red giant can no longer generate the pressures needed to maintain itself against the crushing inward force of its own gravity, and the outer layers begin to fall in toward the center. The red giant collapses.

* The transmutation of heavy hydrogen into helium and heavier elements has been duplicated on the earth for brief moments in the explosion of the hydrogen bomb. However, we have never succeeded in fusing hydrogen nuclei under controlled conditions in such a way that the energy released can be harnessed for constructive purposes. The United States, the Soviet Union and other countries have invested prodigious amounts of money and energy in the effort, for the stakes are high, but physics has not yet been equal to the task. The difficulty is that no furnace has yet been constructed on the earth whose walls can contain a fire at the temperature of the millions of degrees necessary to produce nuclear fusion. The only furnace that can do this is provided by nature in the heart of a star.

At the center of the collapsing star is a core of pure helium, produced by the fusion of hydrogen throughout the star's earlier existence. Helium does not fuse into heavier nuclei at the ordinary stellar temperature of 20 million degrees, because the helium nucleus, with *two* protons, carries a double charge of positive electricity, and, as a consequence, the electrical repulsion between two helium nuclei is stronger than the repulsion between two protons. A temperature of 200 million degrees is required to produce collisions sufficiently violent to pierce the electrical barrier between helium nuclei.

As the star collapses, however, heat is liberated and its temperature rises. Eventually the temperature at the center reaches the critical value of 200 million degrees. At that point helium nuclei commence to fuse in groups of three to form carbon nuclei, releasing nuclear energy in the process and rekindling the fire at the center of the star. The additional release of energy halts the gravitational collapse of the star. It has obtained a new lease on life by burning helium nuclei to produce carbon.

In stars the size of the sun, the helium-burning stage lasts for about one hundred million years. At the end of that time the reserves of fuel, composed now of helium rather than hydrogen, once again are exhausted, and the center of the star is filled with a residue of carbon nuclei. These nuclei, possessing *six* positive electrical charges, are separated by an even more formidable electrical barrier than helium nuclei, and collisions of even greater violence are required for its penetration. The 200-million-degree temperatures which fuse helium nuclei are not adequate for the fusion of carbon nuclei; no less than 600 million degrees are required.

Since the temperatures prevailing within the red giant fall short of 600 million degrees, the nuclear fires die down as the carbon accumulates, and the star, once again lacking the resources needed to sustain it against the weight of its outer layers, commences to collapse a second time under the force of gravity.

All stars lead similar lives up to this point, but their subsequent evolution and manner of dying depend on their size and mass. The small stars shrivel up and fade away, while the large ones disappear in a gigantic explosion. The sun happens to lie just below the dividing line; we are not certain which turn it will take at the end of its life, but we suspect that it will fade away.

The paths of the small and large stars diverge because of differences in the amount of heat generated during the second collapse, at the end of the red giant stage. In a small star the collapse generates a modest amount of heat, and the temperature at the center fails to reach the 600-million-degree level required for the ignition of carbon nuclei. Thus the nuclear fire is never rekindled. Instead the collapse continues, until finally the matter within the star is so compressed that it resists a further reduction in size. The star then remains forever in this highly compressed state. Roughly the size of the earth, it has been squeezed by the force of its own weight into a space only a millionth as large as the volume it originally occupied. A teaspoon of matter from the center of this compact body would weigh 10 tons. If we should ever come across such a star, even though its surface temperature might have declined to a comfortable level, we would be unable to land on this strange world, because its gravity would crush a visitor with a force of 100 million pounds.

Although the center of the star never gets hot enough to burn carbon, the temperature of the surface rises sufficiently so that the star appears white-hot to the eye. These shrunken white-hot stars are called *white dwarfs*. Slowly the white dwarf radiates into space the last of its heat. In the end its temperature drops, and it fades into a blackened corpse.

A very different fate awaits a large, massive star. Because the weight of the star is so great, its collapse generates an enormous amount of heat, greater than the heat generated in the creation of the white dwarf. Soon the temperature reaches the critical level of 600 million degrees at which carbon nuclei fuse together. The fusion of the carbon nuclei forms still heavier elements, ranging from oxygen up to sodium.

Eventually, the carbon fuel reserves are also exhausted; once again their exhaustion is followed by further stages of collapse, heating, and renewed nuclear burning, leading to the production of still other elements.

In this way, through the alternation of collapse and nuclear burning, a massive star successively manufactures all elements up to iron. But iron is a very special element. This metal, which lies halfway between the lightest and heaviest elements, has an exceptionally compact nucleus, whose neutrons and protons are so tightly packed that no energy can be squeezed from it in any sort of nuclear reaction. In fact, nuclear reactions in iron absorb energy; they have the same effect as water thrown on hot coals. When a large amount of iron accumulates at the center of the star, the fire cannot be rekindled; it goes out for the last time, and the star commences a final collapse under the force of its own weight.

The ultimate collapse is a catastrophic event. The iron nuclei at the center soak up the energy of the star as fast as it is generated, and the collapsing materials, meeting negligible resistance, fall in toward the center at enormous speeds, covering a million miles in less than a minute. They pile up at the center, in a dense lump, creating enormous pressures. When the pressure at the center is sufficiently great, the collapse comes to a halt. The collapsed star, compressed like a spring, is momentarily still—and then it rebounds in a violent explosion.

Temperatures ranging up to trillions of degrees are generated during the collapse and subsequent explosion. At these temperatures some of the nuclei in the exploding star are disintegrated, and many neutrons are freed. The neutrons are captured by other nuclei, building up the heavier elements, such as silver, gold, and uranium. In this way the remaining elements of the periodic table, extending beyond iron, are manufactured in the final moments of the star's life.

The explosion hurls out to space all the elements that the star has been manufacturing during its lifetime, leaving only the small, dimly glowing core. The entire episode lasts a few minutes, from the onset of the collapse to the final explosion. This is a short interval for the demise of an object which may have lived for a million years.

The exploding star is called a supernova. Supernovas blaze up with a brilliance many billions of times greater than the brightness of the sun; if the supernova happens to be nearby in our Galaxy, it appears suddenly as a new star in the sky, brighter than any other, and easily visible with the naked eye in the daytime. The last supernova seen in Europe exploded in 1572 and caused a sensation. One of the earliest reported supernovas was a brilliant explosion recorded by Chinese astronomers in A.D. 1054. At the position of this supernova there is today a great cloud of gas known as the Crab Nebula, expanding outward at a speed of seven hundred miles per second, which contains the remains of the star that exploded nine hundred years ago.

What happens to the compressed core of the supernova after its outer layers explode into space? The answer to this question was unknown until 1967. In that year, pulsars—the most interesting objects to be found in the sky for many years—were discovered.

The discovery came about by pure chance. Jocelyn Bell, an astronomy student at Cambridge University, was assigned the task of investigating fluctuations in the strength of radio waves from distant galaxies. She found unexpectedly that certain places in the heavens were emitting short, rapid bursts of radio waves at regular intervals. Each burst lasted no more than one hundredth of a second. The rapid succession of bursts seemed like a speeded-up, celestial Morse code.

The interval between successive bursts was extraordinarily constant. In fact, it did not change by more than one part in 10 million. A clock with this precision would gain or lose no more than a second a year.

No star or galaxy had ever before been observed to emit signals as bizarre as these. At first, some astronomers thought that intelligent beings on other stars might be beaming a message to the earth, and referred to the Morse-code stars as LGM's, standing for Little Green Men. But it soon became evident that the radio pulses had a natural and not an artificial origin. One of the main reasons for this conclusion was the fact that the signals were spread over a broad band of frequencies. If an extraterrestrial society were trying to signal other solar systems, its interstellar transmitters would require enormous power in order for the signals to carry across the trillions of miles that separate every star from its neighbors. The only feasible way to do this would be to concentrate all available power at one frequency, as we do when we broadcast radio and TV programs. It would be wasteful, purposeless and unintelligent to diffuse the power of the transmitter over a broad band of frequencies.

This cold reasoning dashed the hopes of romantics who believed for a short time that man might be receiving his first message from outer space. "LGM's" disappeared from scientific conversation, "pulsars" took their place, and scientists settled down to a search for a natural explanation of the peculiar signals.

The first clue to the answer was the sharpness of the pulses. From the fact that each pulse lasted for one hundredth of a second or less, astronomers concluded that a pulsar is an incredibly small object for a star, far smaller than a white dwarf. This conclusion was based on the fact that when an object emits a burst of radio waves, the waves from different parts of the object arrive at the earth at different times, blurring the sharpness of the original pulse. The smaller the object, the sharper the pulse. Following this line of reasoning, the astronomers calculated that the objects were no more than 10 miles in radius.

This is a startling conclusion. Until then the white dwarf—about 10,000 miles in radius—was thought to be the smallest, densest star in the universe. How could a star be a thousand times smaller than a white dwarf? The matter in such a star would be a billion times denser than the matter in a white dwarf. If the entire earth were compressed to the same degree as a pulsar, it would fit into the Pentagon. If the Pentagon were compressed as much, it would be the size of a pinhead.

The answer goes back to a prediction made several decades ago. At that time, several theoretical astronomers pointed out that when a large star collapses at the end of its life, just before it explodes as a supernova, the materials of the collapsing star pile up at the center, and produce enormous pressures, even more powerful than the inward pressure produced by the star's own weight. Under this crushing burden, the individual electrons and protons within the star are forced to combine into neutrons. A pure ball of neutrons forms at the center of the star, only 10 miles in radius, but with most of the star's original mass packed into it. The hypothetical ball of neutrons was dubbed a "neutron star."

Starting in 1965, astronomers searched for neutron stars assiduously, investigating with particular care the region at the center of the Crab Nebula, where the squeezed-down core of the supernova explosion of A.D. 1054 should have been located. But no neutron stars were discovered, and interest in them faded.

In 1968 a wave of excitement spread through the astronomical community when a pulsar was discovered at the center of the Crab Nebula, at precisely the place where they had previously searched for a neutron star. Suddenly, many items of evidence fitted together like the pieces of a jigsaw puzzle: a neutron star was predicted to exist at the center of the Crab Nebula; a pulsar was discovered at the center of the Crab Nebula; and the neutron star and the pulsar are the only objects known that have the mass of a star packed into a 10-mile sphere. Clearly, neutron star and pulsar were two names for the same thing: a fantastically compressed, super-dense ball of matter, created when a massive star collapses at the end of its life.

One mystery remained to be explained. What produces the sharp, regularly repeated bursts of radiation from which pulsars derive their name? The answer is believed to be that a pulsar, like the sun and most other stars, is subject to violent storms which may last for years, spraying particles and radiation out into space. Each storm occurs in a localized area on the surface of the pulsar and sprays its radiation into space in a narrowly defined direction. When the earth lies in the path of one of these streams of radiation, our radio telescopes pick up the signals, which indicate to us the presence of the pulsar.

But if the spray of radiation is emitted steadily from the pulsar, why do we observe it as a succession of isolated sharp bursts? The reason is probably that pulsars, like most stars, spin on their axes. In fact, being smaller than normal stars, pulsars can spin very rapidly, as many as several times a second. As the pulsar spins, the stream of radiation from its surface sweeps through space like the light from a revolving lighthouse beacon. If the earth happens to lie in the path of the rotating beam, it will receive a sharp burst of radiation once in every turn of the pulsar.

This theory can be checked, because all spinning objects slow down steadily in the course of time as a result of friction. Thus the interval of time between successive bursts of radiation from a pulsar must increase. In 1969 this prediction was confirmed by the discovery that the time between successive pulses from the Crab Nebula pulsar was getting longer, at the tiny but measurable rate of one-billionth of a second per day.

With the realization of the connection between neutron stars, pulsars and supernovas, many astronomers feel that the final pages may have been written in the life story of the stars. But others suspect that at least one surprise is still in store for us, for there is reason to believe that the neutron star or pulsar is not the ultimate state of compression of stellar matter. Under certain conditions, a star may continue to collapse beyond the 10-mile limit of the neutron star, falling inward upon itself faster and faster, until it has contracted to a radius of two miles. At this point, the theory of relativity predicts the occurrence of an extraordinary phenomenon.

According to Einstein's theory, energy and mass are equivalent. The equivalence is expressed in the famous equation

$$E = mc^2$$

where E is the energy of the light ray, m is its mass and c is the velocity of light. Reference is frequently made to this equation in calculating the energy E yielded by the annihilation of an amount m of uranium in the explosion of a nuclear bomb. What has this to do with the collapsing star? If energy is equivalent to mass, a ray of light which possesses electromagnetic energy must also possess mass, just as if it were a particle of matter. Thus, a ray of light emitted from a star will be pulled back by the star's gravity, as a ball thrown up from the surface of the earth is pulled back by the earth's gravity. When the star is normal in size—about a million miles in diameter—the force of gravity on its surface is not strong enough to keep the light rays from escaping, and they leave the star, although with somewhat less energy. But as the star contracts, the force of its gravity grows rapidly, and by the time the star's diameter has decreased to four miles, the gravity at its surface is billions of times stronger than the force of gravity on the surface of the sun. The tug of this enormous force prevents the rays of light from leaving the surface of the star; like the ball thrown upward, they are pulled back and cannot escape to space. From this moment on, the star is invisible. It is a *black hole* in space.

Inside the black hole, the contraction continues, piling up matter at the center in a tiny, incredibly dense lump. According to current knowledge in theoretical physics, this is the end of the star's life. The star's volume becomes smaller and smaller; from a globe with a two-mile radius it shrinks to the size of a pinhead, then to the size of a microbe, and, still shrinking, passes into the realm of distances smaller than any ever probed by man. At all times the star's mass of a thousand trillion trillion tons remains packed into the shrinking volume. But intuition tells us that such an object cannot exist. At some point the collapse must be halted. Yet, according to the laws of twentieth-century physics, no force, no matter how powerful, can stop the collapse. The implication is that the laws of physics must be modified at extremely short distances in a manner that prevents particles from coming infinitely close together. Here is a hint of the impending discovery of a new law or a new agent in nature, one that may some day lead to the liberation of energies even greater than the energy of the nucleus. Such a discovery would transform the world of the future as the discovery of the nuclear force transformed the world of the twentieth century. The study of the stars may yet control the affairs of men.

The life story of the stars has an epilogue. When a supernova explosion occurs and the outer layers of the stars are sprayed out to space, they mingle with fresh hydrogen to form a gaseous mixture containing all 92 elements. Later in the history of the galaxy, other stars are formed out of clouds of hydrogen which have been enriched by the products of these explosions. The sun is one of these stars; it contains the debris of countless supernova explosions dating back to the earliest years of the Galaxy. The planets also contain the debris; and the earth, in particular, is composed almost entirely of it. We owe our corporeal existence to events that took place billions of years ago, in stars that lived and died long before the solar system came into being.

A CLOUD PULLED TOGETHER BY GRAVITY. It is believed that stars are born in the swirling clouds of hydrogen gas which fill all of space. If the atoms in one region of such a cloud come together by accident or are forced together by the pressure of the surrounding clouds, the force of gravity pulls the atoms still closer together, forming a condensed pocket of gas *(below, left)*. The continuing action of gravity compresses the pocket of gas *(below right)*. As a result of the compression, the temperature at the center rises; after about 10 million years, when the temperature has reached the critical value of 20 million degrees, nuclear reactions set in, in which vast amounts of energy are released. The onset of these reactions marks the birth of a star. The release of nuclear energy halts the further collapse of the star. The energy passes to the surface and radiates into space in the form of light and heat.

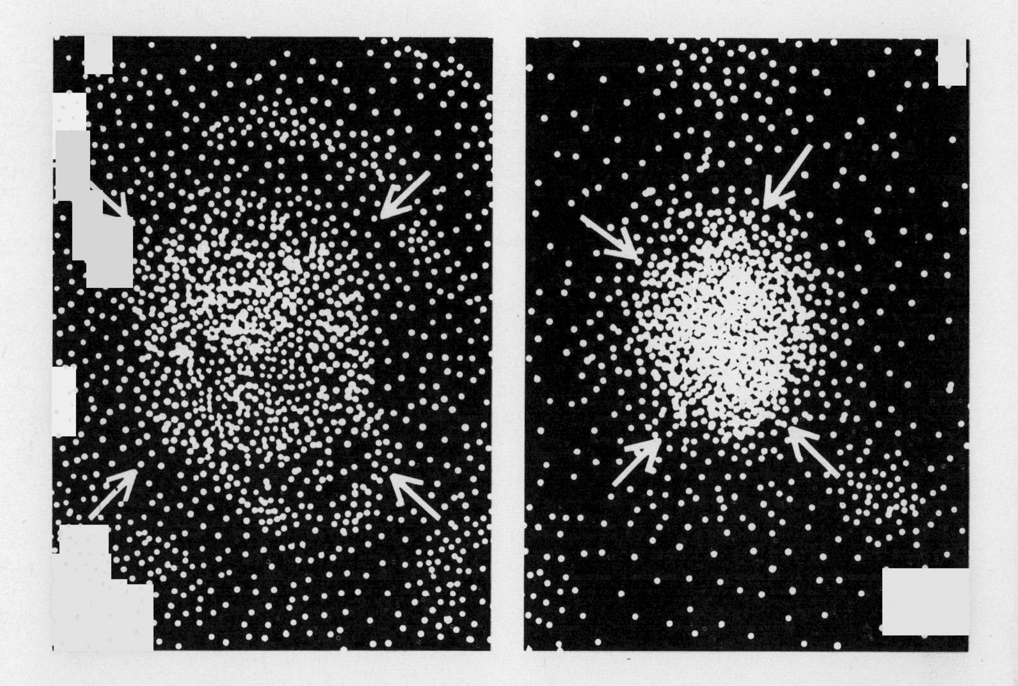

The photograph opposite shows a dense cloud of gas and dust in which stars are being born, situated about 4000 light-years from the sun in the Constellation Unicorn. The luminous regions are clouds of hydrogen sighted by ultraviolet radiation from nearby stars. The dark areas are opaque clouds of dust.

STARS IN FORMATION

THE MANUFACTURE OF ELEMENTS WITHIN STARS. Throughout most of the life of a star its energy is derived from the fusion of hydrogen nuclei, or protons, into helium nuclei. Four protons unite to form one helium nucleus in this reaction, shedding, at the same time, two units of positive electric charge in the form of positrons accompanied by neutrinos *(below)*.

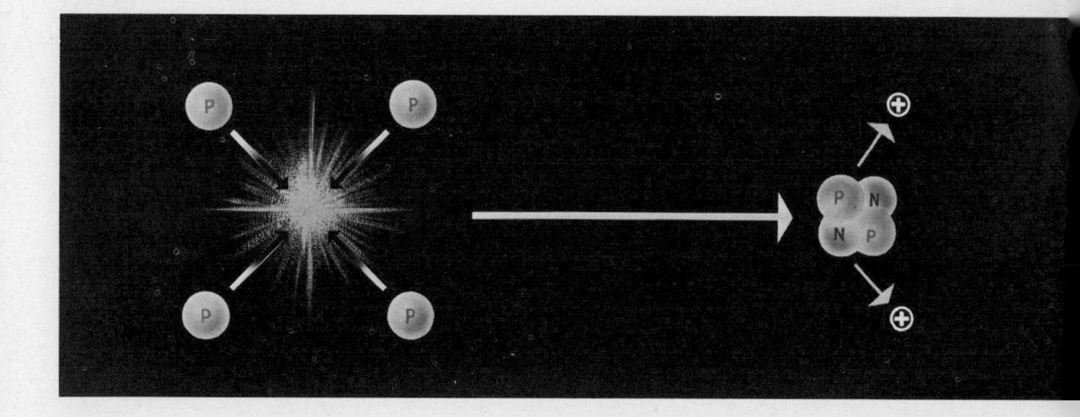

The fusion of hydrogen to form helium continues for 99 percent of a star's lifetime. When the hydrogen is substantially depleted the star again collapses until its center reaches a temperature of 200 million degrees Fahrenheit. At this temperature helium nuclei fuse to form the nucleus of carbon *(below)*. Oxygen and still heavier elements are formed after carbon. In this way, the elements of the universe are synthesized, step by step, out of the basic building block of hydrogen.

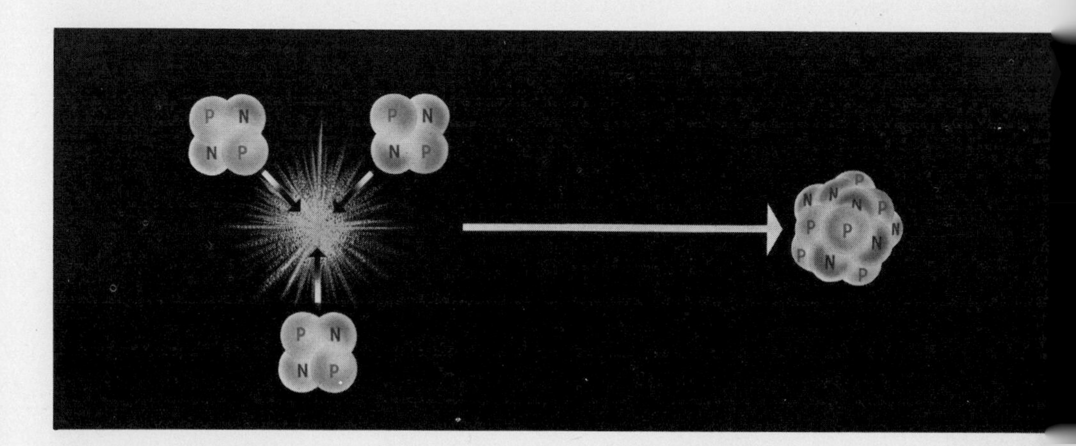

THE DEATH OF A STAR. At the end of a star's life, when its nuclear energy is spent, the star collapses under the force of its own weight. In the case of a small star the collapse continues until the entire mass is squeezed into a volume the size of the earth. Such highly compressed stars, called white dwarfs, have a density of ten tons per cubic inch. Slowly the white dwarf radiates into space the last of its heat and fades into darkness.

A different fate awaits a large star. Its final collapse is a catastrophic event which generates temperatures of several billion degrees, burning the last residue of fuel sprinkled throughout the star, and releasing a burst of energy which blows the star apart. The exploding star is called a supernova. Supernovas may be 10 billion times brighter than the sun. If the supernova is located nearby in our galaxy, it appears suddenly as a very bright star, visible in the daytime sky.

One of the earliest reported supernovas was observed by Chinese astronomers in A.D. 1054. Today at the position of this supernova there is a large cloud of gas known as the Crab Nebula, shown in the photograph above, which is expanding outward at a speed of 1000 miles per second.

The supernova explosion sprays the material of the star out into space, where it mingles with fresh hydrogen to form a mixture containing all 92 elements. Later in the history of the galaxy, other stars are formed out of clouds of hydrogen which have been enriched by the products of these explosions. The sun, the earth, and the beings on its surface—all were formed out of such clouds containing the debris of supernova explosions dating back billions of years to the beginning of the Galaxy.

ASTRONOMERS DISCUSS UNSOLVED PROBLEMS. Great progress has been made in the past ten years in the study of stellar evolution, but uncertainties remain. In January, 1965, a conference on stellar evolution was held at the Goddard Institute for Space Studies. In this conference Chushiro Hayashi of the University of Tokyo reported the results of calculations on the first stage of a star's life, in which it is contracting under the force of its gravity but has not yet heated up to the point at which hydrogen burning begins. Dr. Hayashi discovered, in contrast to earlier predictions, that the embryonic star is hundreds of times more luminous, while still contracting, than it is after nuclear burning

ASTRONOMERS DISCUSS UNSOLVED PROBLEMS

commences. The source of this vast outpouring of energy is to be found in gravitational forces within the star. Dr. Hayashi's calculations, which have subsequently been confirmed by other astrophysicists, have important consequences for theories of the early history of the solar system.

Standing at the right of Dr. Hayashi in the photograph below is Geoffrey Burbidge, chairman of this session of the conference. Seated to his right is Bengt Strömgren, recipient of the Gold Medal of the Royal Astronomical Society for his research on stellar evolution and on the nature of the gas and dust in the space between stars.

☼

5 The Beginning and the End

A STAR is born out of the condensation of clouds of gaseous hydrogen in outer space. As gravity pulls together the atoms of the cloud, its temperature rises until the hydrogen nuclei within it begin to fuse and burn in a series of reactions, forming helium first, and then all the remaining substances of the universe. The elements of which our bodies are composed were manufactured in this way, in the interiors of stars now deceased, and distributed to space when these stars exploded. Subsequently, these elements were drawn together again in the cloud of gas out of which the sun and the earth condensed. If the sun explodes at the end of its life the planets will be consumed, and their substance once again distributed into space, to be reincarnated in another solar system as yet unborn.

This beautiful theory allows the universe to go on forever in a timeless cycle of death and rebirth, but for one disturbing fact. Fresh hydrogen is the essential ingredient in the plan, but with the passage of time the supply of fresh hydrogen dwindles; as the old stars go out, one by one, fewer and fewer new stars are formed to replace them. Stars are the source of the energy by which all beings live. When the light of the last star is extinguished, life must end throughout the universe.

Other evidence suggests that the universe is changing in an irreversible way. All the galaxies seem to be moving away from one another at very high speeds. Those most distant from us are receding at the extraordinary speed of 150,000 miles per second, which is close to the velocity of light. The universe appears to be blowing up before our eyes, as if we were witnessing the aftermath of a gigantic explosion. As the galaxies fly apart and the distances between them increase, space grows emptier, and the density of matter dwindles to nothing. It seems, that, albeit slowly, the universe is approaching an end.

Some years ago, Thomas Gold, then a graduate student at Cambridge University, made a proposal which voids these morbid predictions. He suggested that fresh hydrogen is steadily created throughout the universe *out of nothing.* The freshly created hydrogen would provide the ingredients for the formation of new stars to replace the old. At the same time it would fill up the spaces left by the movement of the galaxies away from one another. Thus, the creation of matter out of nothing, as proposed by Gold, could restore the universe to a state of perpetual balance, without beginning and without end.

Gold mentioned his idea to Herman Bondi and Fred Hoyle, two English astronomers, who joined him in working out its consequences. They asked themselves, how much hydrogen should be created per year in order to keep the density of matter constant everywhere, as the universe expands? According to their calculations, the expanding universe remains in a steady state, with a constant density of matter, if one hydrogen atom is created per year in a volume equal to that of the Empire State Building.

This is a very modest rate of creation, but it violates a cherished concept in science—the principle of the conservation of matter—which states that matter can be neither created nor destroyed. It seems difficult to accept a theory that ignores such a firmly established fact of terrestrial experience. Yet, the proposal for the creation of matter out of nothing possesses a strong appeal, for it permits us to contemplate a universe that extends into the past and the future without limit, a universe that renews itself *in perpetuum.*

Another cosmology, which offers strong competition to the steady state theory, makes no attempt to dodge the implications of the expanding universe. This cosmology, appropriately named the big-bang theory, proposes that the expansion is in fact the consequence of an actual explosion which took place a long time ago. Father Lemaître, a Belgian astronomer educated as a Jesuit priest, and George Gamow, a Russian-born physicist who emigrated to the United States in 1936, are the scientists most prominently associated with this theory. In 1931 Lemaître proposed that the universe began its existence as a condensed droplet of matter at an extremely high density and temperature. Later Gamow named this primordial egg "ylem"—the name that Aristotle gave to the basic substance out of which the ancients believed all matter was derived. Internal pressures within this hot, dense droplet, containing all of the matter and radiation of the universe, caused it to expand rapidly. As it expanded, its temperature and pressure dropped. In the first few minutes of its existence the temperature was many millions of degrees, and all the matter within the droplet consisted of the basic particles—electrons, neutrons and protons. Any combination of these particles to form nuclei or atoms would quickly be disintegrated under the smashing impact of the violent collisions which occur at such high temperatures. However, as the universe continued its expansion, and the temperature of ylem dropped further, the protons and neutrons began to fuse together into nuclei, and remained fused together for increasingly long periods of time. They formed, first, deuterium, then helium, and then still heavier elements. According to the big-bang theory of Lemaître and Gamow, all 92 elements were formed in this way in the first half-hour of the existence of the universe.

With the further passage of time, the matter of the universe cooled and condensed into galaxies, and within the galaxies, into stars. After some billions of years of continued expansion, the universe reached the state in which it exists today. Knowing how far apart the galaxies now are, and how rapidly they are moving away from one another, we can calculate backward in time to the moment at which the expansion began. In this way the advocates of the big-bang cosmology arrived at the conclusion that the universe began its existence 10 billion years ago. At the present time, in the words of Lemaître,

> the evolution of the world can be compared to a display of fireworks that has just ended: some few red wisps, ashes, and smoke. Standing on a cooled cinder, we see the slow fading of the suns, and we try to recall the vanished brilliance of the origin of the worlds.

Which cosmology is correct? Was there a beginning? Will there be an end? In 1965 a discovery was made which throws some light on this subject. In that year Professor Robert Dicke pointed out an aspect of the big bang which seemed to have escaped notice. According to the big-bang theory, the universe began as a droplet of hot, dense matter. This early universe must have been a fireball filled with an intense brilliant radiation. As the universe expanded, the intensity of the radiation diminished. But Dicke's calculations indicated that a remnant of the fireball radiation should exist today, and should be detectable with a sensitive radio antenna.*

Dicke set about constructing an apparatus to search for the remnant of the primordial fireball radiation, unaware that two Bell Laboratory physicists—Drs. Arno Penzias and Robert Wilson—had already discovered it. They, too, were unaware that they had made the discovery, for they were not looking for fireball radiation; they were measuring the intensity of radio noise received in a large antenna which had been set up some time before in connection with the communications satellite program. Their measurements revealed a puzzling radiation which came to their antenna from all parts of the sky. Penzias

* Gamow had the same idea ten years earlier, and presented it in an article published in 1956, but his suggestion drew no reaction at that time.

and Wilson were unable to explain the source of this radiation, until a friend told them of Dicke's work. The rest is scientific history.

Subsequently, other physicists and astronomers confirmed the existence of the primordial fireball radiation. Their measurements constitute strong evidence for the big-bang cosmology. No competing explanation for this radiation has been provided thus far by the steady-state cosmologists.

Thus, the facts seem to favor the big-bang cosmology. If this theory is correct, the universe began suddenly, some 10 billion years ago. But what are we to make of such a picture? The universe is the totality of matter; if there was a beginning, what came before? When all the stars go out, what comes after? In a desperate effort to secure an infinite lifetime for the universe, some cosmologists say that we may be oscillating back and forth forever between our condensed and expanded states. However, there is evidence against this oscillating theory.

There the matter rests for the moment. Scientists have exposed very interesting features of the Great Plan—the birth of stars, the assemblage of the elements within the stars out of the three basic particles, and their dispersal to space in supernova explosions; but science offers no satisfactory answer to one of the most profound questions to occupy the mind of man—the question of beginning and end.

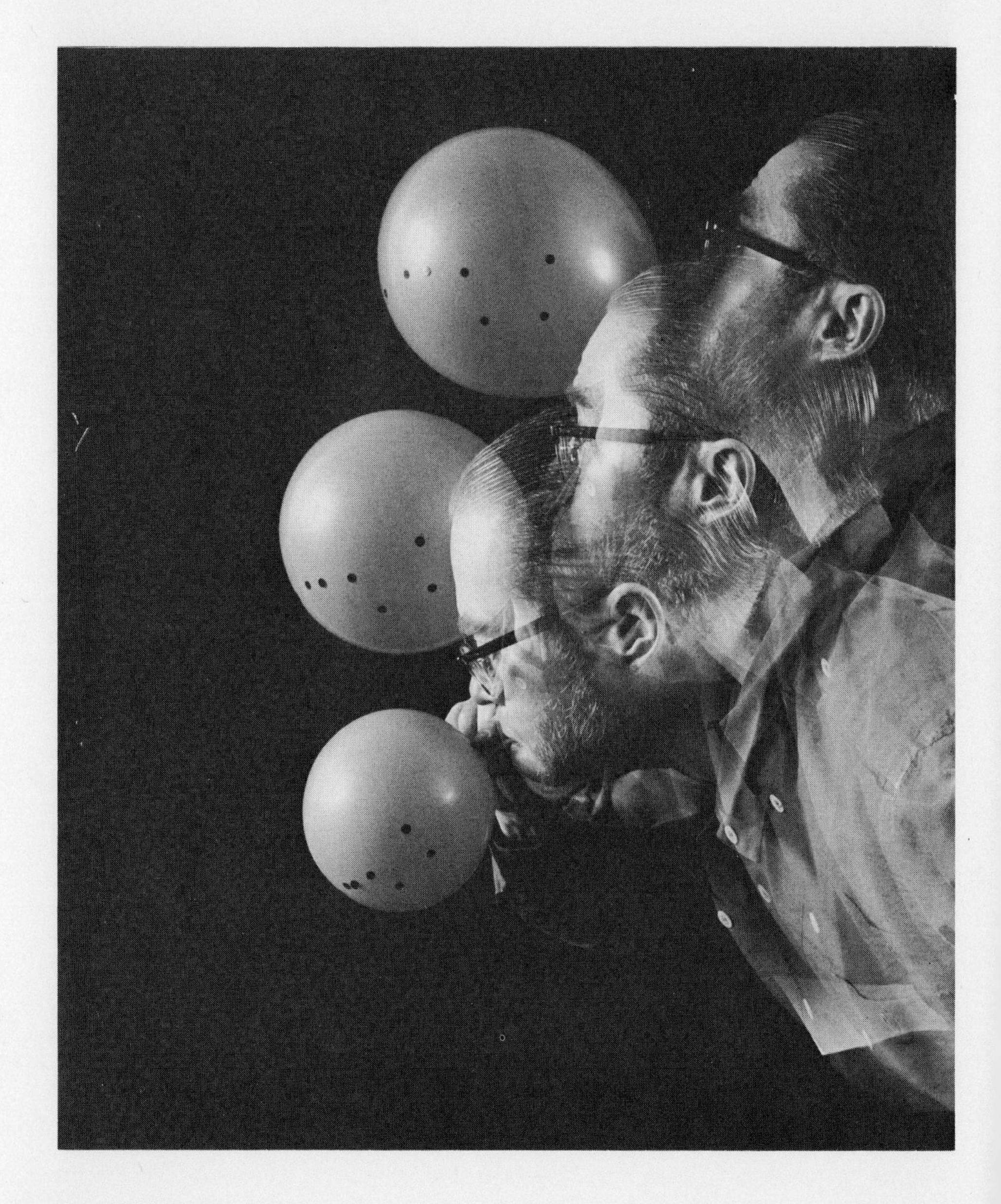

THE EXPANDING UNIVERSE. Astronomical observations indicate that all the galaxies in the universe are moving away from one another at very high speeds, just as if they were on the surface of a rapidly expanding balloon *(above)*. The nature of the expansion is difficult to conceive when it occurs in the three dimensions of the real world rather than on the two-dimensional surface of a balloon, but the principle is the same.

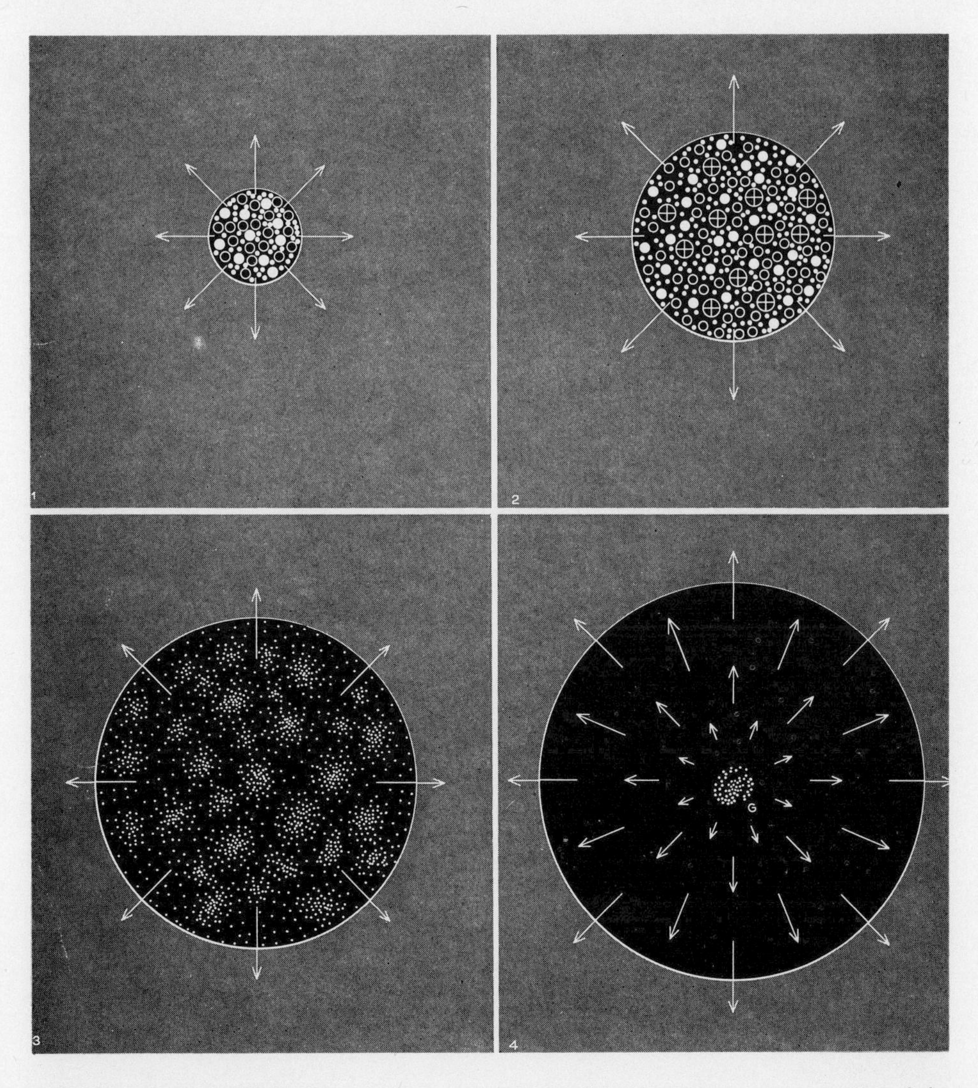

THE BIG-BANG COSMOLOGY. In an attempt to explain the apparent expansion of the universe, George Gamow proposed that 10 billion years ago the universe was a hot, dense cloud into which the basic particles of matter—electrons, neutrons and protons—were packed at exceedingly close quarters and a very high temperature. The highly compressed universe expanded outward with explosive violence. The first figure shows the universe a few minutes after the initial explosion, when the temperature is several million degrees. In the second figure, one hour later, the universe is still expanding, but the temperature has dropped; as a consequence, neutrons and protons are combining to form nuclei. In the third figure, 10 million years later, the temperature has fallen to 1000 degrees, and atoms have formed and have started to condense into galaxies, and into stars within the galaxies. The last figure shows one of these galaxies, appearing from this viewpoint to be at the center of the expansion because all neighboring galaxies are moving away from it.

THE STEADY-STATE COSMOLOGISTS. Another effort to explain the expansion of the universe was made by the English cosmologists, Herman Bondi, Tom Gold and Fred Hoyle. Their theory, called the steady-state cosmology, attempts to avoid the philosophical objections to a theory in which the universe has a beginning at a definite point of time, by proposing that hydrogen is continuously created out of nothing throughout the universe. They suggested that the newly created hydrogen would fill the void left by the expansion

of the universe and, at the same time, provide fresh material for the formation of stars. Thus the universe renews itself indefinitely, without beginning and without end.

In the photograph Hoyle *(left)* and Gold *(middle)* are shown engaged in conversation with Freeman Dyson of the Institute for Advanced Study in Princeton, at a conference held at the Goddard Institute for Space Studies in 1961.

THE PRIMORDIAL FIREBALL. Strong evidence in support of the big-bang cosmology appeared in 1965, when Robert Dicke of Princeton University pointed out that according to this theory the universe began as a dense, hot fireball, filled with intense radiation. He predicted that a remnant of the primordial fireball radiation should be detectable today with sensitive radio antennas. As he began to construct an apparatus to search for the fireball radiation, two Bell Laboratory physicists, Arno Penzias and Robert Wilson, stumbled across it while measuring the radio signals received by a large antenna set up by A.T.&T. for the communications satellite program.

Robert Dicke

Arno Penzias *(right)* and Robert Wilson in front of the Holmdel antenna.

☼

6 The Origin of the Solar System

THE origin of the solar system is less of a mystery than the origin of the universe, but it is still not a clearly understood event. When I was in high school I was taught the theory that the planets came into being as by-products of a catastrophic event in which the sun collided with a passing star. The force of gravity tore huge streamers of flaming gas out of the bodies of the two stars during this encounter. As the intruding star receded into the distance, some of these streamers of gaseous material were attracted by the sun's gravity and captured into orbits circling around it. The earth condensed out of one of these streams of hot gas to form a molten mass, on whose surface a crust formed and gradually hardened with the passage of time.

It is easy to calculate the probability that the solar system originated in this way. The likelihood of a collision between the sun and another star depends on the size of the sun, and on the distance between it and its neighbors. Stars, large though they are, are very small in comparison with the average distances that separate them. The sun, for example, is one million miles in diameter, and 20 trillion miles from its nearest neighbor. For that reason, the chance of a collision between two stars is very small; a calculation shows, in fact, that one collision, or at most a few, may have occurred during the history of our galaxy. Indeed, the planets in our solar system may be the only ones in existence, according to the collision theory. It is well-nigh impossible that two of these rare objects—planets—should be located next door to one another on neighboring stars.

Yet, a few years ago a planet was found orbiting around one of our nearest neighbors in the sky, called Barnard's star. The newly discovered planet was detected by its influence on the motion of its parent star. As a planet revolves around a star, the pull of its gravity causes the star to follow the motion of the planet in its orbit to some degree. When the star is observed from the earth through a telescope we can see it move back and forth as the planet circles around it. Because the planet is so much lighter than the star, its pull on the latter is quite weak, and the change in the star's position is correspondingly slight, but it can be detected if the star is sufficiently close to us. In the case of Barnard's star the change was exceedingly minute, being equivalent to a displacement of a hair's width viewed from a distance of one mile, and twenty-five years of observation were required to confirm its existence.

Barnard's star is located only six light years away from the sun, which is a stone's throw in the scale of stellar distances. The discovery of the planet circling Barnard's star casts doubt on the validity of the collision theory, because it contradicts the prediction of the theory that stars with planets circling around them should be an unusual phenomenon, spaced very far apart in the Galaxy.

The collision theory suffers from an additional weakness. It is unable to account for the fact that the orbits of most of the planets are nearly perfect circles. The orbit of the earth, for example, departs from a perfect circle by only 2 percent, and the orbit of Venus departs from one by only seven-tenths of 1 percent. According to the collision theory the orbits of the planets should be elongated and narrow, because the planets were formed from long narrow filaments of gas stretching out toward the retreating star. It is possible that the narrow orbits predicted by the collision theory could have become changed or transformed into perfect circles over the course of billions of years, but no one has been able to think of a natural way in which this change could have come about.

These discrepancies force us to look elsewhere for a theory of the origin of the solar system. Fortunately, a competing theory lies close at hand. The proponents of the second theory suggest that planets are formed in a manner similar to the formation of the sun itself, as small condensations in the cloud of partly compressed gas which surrounds the sun at the time of its birth. As the gas cloud of the sun-to-be contracts under the inward force of its own gravity, it leaves behind at its outer edge all those atoms of gas which are moving fast enough so that their forward motion, combined with the inward motion toward the center of the cloud under the attraction of gravity, curves their paths into circular or near-circular orbits. Each of these atoms of gas is bound to the central gas cloud by gravity and revolves around it like a miniature planet. The central cloud continues to contract until its temperature reaches the critical level of 20 million degrees needed for the ignition of thermonuclear reactions. This point in the contraction signals the birth of the sun. At the same time the surrounding gas atoms, circling in orbits around the sun, condense under the influence of gravity to form the planets. Presumably the moon and the other natural satellites of the planets were formed in the same manner as still smaller condensations around their parent planets.

If stars and planets are both formed by the condensation of a cloud of gas under gravity, what is the difference between them? The answer is connected with their masses. As the star-cloud, or the planet-cloud, contracts, its temperature rises. If the cloud is very massive the temperature reaches the value of 20 million degrees, at which nuclear burning commences, and a star is born. If the cloud is small, the temperature fails to reach the critical level, and the condensed cloud remains an inert body without internal sources of nuclear energy; that is, it becomes a planet.

The smallest cloud of gas that will create a star is approximately one-tenth the mass of our sun. Jupiter is a very large planet, 318 times more massive than the earth, but still it falls short, by a factor of about 100, of having the mass needed to form a second star in our solar system.

It is possible that some of the clouds of gas which condense in outer space are too small to produce the temperatures needed for nuclear burning and, therefore, cannot become stars; yet they cannot properly be called planets, because a planet is a body that is bound to a star by gravity and circles in orbit around it. Such relatively cold, planet-sized bodies, free of the gravitational influence of any star, may exist around us in considerable numbers. Since they do not shine by their own light, and are too far away from any star to be visible to us by its reflected light, we have no way of detecting these "free planets" at present. We may stumble across one of them when we begin to explore with space vehicles the regions lying beyond the limits of the solar system.

These thoughts are derived from the condensation theory of the origin of the solar system, for which no definite proof has been supplied. However, there are good reasons for believing in the theory. First, it fits naturally into the latest ideas on the birth of stars. Second, it predicts that the planets are a natural accompaniment to the birth of stars, and the prediction is borne out by the detection of planets circling around stars in the neighborhood of the sun. Third, it offers a natural explanation of the circular orbits of the planets in our solar system.

Nonetheless, the theory presents some difficulties. Foremost among these is the problem of the sun's rotation. The sun spins on its axis once every 27 days, just as the earth rotates once in 24 hours. It is easy to understand why the sun should be spinning in this way if we remember that it was originally formed out of swirling masses of interstellar hydrogen gas. Some of the swirling motion must have been retained in the cloud out of which the sun formed. When the sun-cloud was newly formed, and its dimensions were large, it probably rotated very slowly. But as it contracted it must have spun more and more rapidly, just as an ice skater spins faster when he pulls in his arms and draws his skates together, in contracting from a sweeping turn to a small circle. As a result of this effect the sun should now be spinning on its axis at the rate of once every few hours. Actually, it turns at a far slower rate, 100 times less rapid. What has slowed the sun down? A thoroughly satisfactory answer has never been provided.

Another problem arises out of the fact that the planets should, according to theory, contain a large fraction of the mass of the original solar gas cloud. A calculation by James Jeans leads to the conclusion that a third of the mass of the sun should have been left behind to form the planets. Actually, the planets have little more than a thousandth of the mass of the sun. What has happened to the missing material? Perhaps streams of energetic particles from the surface of the sun blasted it away, or it may have evaporated from the outer regions of the solar system. The answer is not known.

These difficulties of the condensation theory are not easily resolved. It can only be said that we have a suspicion that the earth was formed by condensation, along with the sun, 4.5 billion years ago, but no one yet has a clear understanding of the tangled complex of events that surrounded the genesis of the planets. It is possible that we may be in for a major surprise when we carry out our first *in situ* explorations of the moon and planets in the coming decades.

THE FORMATION OF THE SUN AND PLANETS. The solar system is believed to have begun its life as a diffuse cloud of gas held together by the force of its own gravity (1). The original radius of the cloud was approximately 10 trillion miles.

With the passage of time the attraction of gravity drew the atoms of gas together and the cloud contracted (2).

Over a period of time, which is difficult to determine but was probably between 10 and 20 million years, the dense center of the cloud gradually shrank to approximately the size of the present sun (3). At this point the temperature within the central region was high enough to ignite nuclear reactions, in which hydrogen fused to form helium, releasing energy and marking the birth of the sun.

In the cooler and less dense regions surrounding the primitive sun, smaller condensations developed under the force of gravity, giving rise to the planets (4).

1
2
3
4
MARS
VENUS
SUN
MERCURY
EARTH

METEORITES AND THE ORIGIN OF THE SOLAR SYSTEM. The condensation theory of the origin of the solar system, although widely accepted, nonetheless suffers from several serious weaknesses. In 1961, a conference was held at the Goddard Institute for Space Studies in New York City at which astronomers, physicists and geophysicists gathered to discuss the difficulties in the theory. Above, John Wood of the University of Chicago is shown during a lecture presented to conference participants, in which he discussed meteorites and their significance in understanding the origin of the planets. Meteorites are pieces of extraterrestrial rock moving at high speeds through the solar system like

miniature planets. Occasionally they collide with the earth. The rapid passage of a meteorite through the atmosphere of the earth raises its surface temperature to a white heat; at night the trail of the glowing meteorite is easily visible; it is called a "falling star." Most meteorites are small and burn up in the atmosphere. A few are large enough to reach the ground; once in a great while, one of these is discovered and brought into a laboratory for study. Meteorites are exceptionally interesting because they probably represent the original state of the material out of which the planets were formed; for that reason they are always mentioned in discussions of the origin of the solar system.

☼

7 The Moon: Rosetta Stone of the Planets

WE estimate that the earth and the moon, along with the rest of the solar system, were formed 4.6 billion years ago. Sometime in the first billion years, life appeared on the earth's surface. Slowly, the fossil record indicates, living organisms climbed the ladder from simple to more advanced forms until—perhaps a million years ago—the threshold of intelligence was crossed.

Earthbound creatures can never learn how that happened, or what conditions led to the emergence of life, because the record of the earth's early years has been wiped out. The air and water that make our planet livable have worn down the oldest rocks and washed away their remains into the oceans, while mountain-building activity and volcanic eruptions have churned the surface and flooded it repeatedly with fresh lava, removing the remaining evidence. These natural forces have entirely removed the primitive materials that lay on the earth's surface when it was first formed. No rocks have ever been found on the earth that are older than 3.5 billion years. We know nothing of what happened on our planet from the time of its formation, 4.6 billion years ago, to the time when these oldest rocks were laid down. The critical first billion years, during which life began, are blank pages in the earth's history.

But on the moon there are no oceans and atmosphere to destroy the surface, and there is relatively little of the mountain-building activity that rapidly changes the face of the earth. Over large areas, the materials of the moon's surface are as well preserved as if they had been in cold storage. The moon offers the best chance of recapturing the lost record of the earth's past.

A casual inspection of photographs of the moon immediately confirms that this small planet has retained the record of early events in its history with exceptional fidelity. The photographs show countless craters, most of which have been produced by the impact of meteorites raining down on the moon for billions of years. Many craters are circled by ramparts ranging up to 10,000 feet in height. Some of these ramparts must be a billion years old or more, yet photographs taken with a telescope clearly indicate that they have been preserved almost unchanged, with little of the original material worn away. Meteorites have collided with the earth throughout its history, just as they have collided with the moon, and they have produced similar craters; but all traces of the older craters are gone because on the earth various agents of erosion, of which the most important is running water, move materials from one place to another, leveling the crater walls and filling in the pits and hollows. Only the scars of the most recent collisions, such as the Arizona meteorite craters, are still visible on the earth.

Pictures of the moon taken by early NASA spacecraft provided proof that the extent of erosion on the moon has indeed been very small. The photographs showed lunar features as small as a few feet in diameter. The clarity of the spacecraft photographs produced jubilation among astronomers, who had been straining to see through the earth's atmosphere like drivers peering at the road through a rain-spattered windshield. Many craters which had never been seen before in photographs taken with telescopes on the earth were visible, ranging from a few feet in diameter up to hundreds of feet. These small craters must have existed on the earth as well, but were wiped out almost immediately by the wearing effect of winds and running water; even a 200-foot crater lasts here only a million years or so, which is the blink of an eye in the scale of geologic time. On the moon, such a crater lasts for a billion years, and even the shallow footprints of the Apollo astronauts, no more than six inches deep, will last a million years or more.

Why does the moon lack an atmosphere? The answer is connected with the small size of the moon and the weakness of its gravitational pull. The atmosphere of any moon or planet quickly drifts into space if it is not held at the surface by gravity. Even with gravity there is a steady leakage of gas from the atmosphere into space. The smaller the planet, the less the pull of its gravity and the greater the leakage rate. The moon is so small that all the gases originally present in its atmosphere escaped quickly when it was still very young.

Why does the moon lack water, which is so abundant on the earth? Again the explanation is connected with the weak pull of gravity on the moon. The materials of the moon, like the materials of the earth, probably contained some water 4.6 billion years ago when these bodies first condensed out of the gases of the solar nebula. The molecules of water, trapped in the interior of the moon initially, would have diffused upward from the interior in the course of time. When they reached the surface they would have disappeared, because the moon's gravitational pull was too weak to hold them. During the course of a billion years, or perhaps even sooner, the outer layers of the moon must have become thoroughly dehydrated.

When Galileo, the first man to look at the moon through a telescope, turned his primitive instrument on that body in 1609 he saw large dark areas resembling oceans, which he called *maria,* or seas. The word has remained, but we know now that the resemblance to bodies of water is illusory. The lunar seas contain no water; no storms rage across the plains; no streams flow down from the highlands. The moon is a dry planet.

The dryness of the moon leads us to make a confident prediction; no lunar bugs lie concealed in crevices or under rocks, awaiting discovery by future lunar landing parties. In fact, it is unlikely that any form of life exists on the moon. Water is essential for the development of the kinds of life with which we are familiar, because it provides a fluid medium in which the complex molecules of the cell can travel freely. This free movement leads to frequent collisions between neighboring molecules and, as a consequence of these collisions, the chemical reactions that make up the ongoing processes of life.

Even if all the basic chemicals of the living cell were abundant on the moon and were spread out on its dry surface, they would never unite to form the simplest living organism, because they would be unable to move about.

Water may not be the only fluid that could serve this purpose. Liquid ammonia, for example, might suffice. However, this enlargement of the possibilities for the development of life does not lead to any greater optimism regarding the chances of finding life on the moon. If liquid ammonia were present on the moon, it would escape from the surface as rapidly as water, because the sun's ultraviolet rays break up the molecules of ammonia into their component atoms of hydrogen and nitrogen, and light hydrogen atoms fly off into space. This is true for any liquid we might imagine as being present on the moon in abundance. It is almost certain, then, that the lunar surface is devoid of all forms of life today.

Yet the arid, lifeless moon may reveal clues to the origin of life on the earth. The explanation of this paradox lies in the fact that while the moon is dry today, it may have had some moisture on its surface for a short time in its youth. We are certain that all the molecules of water which reached the surface of the moon from its interior escaped to space fairly rapidly, but it is possible that they remained just long enough for small pools of water to form. In the brief interval during which this moisture was present on the moon, there may also have been present, as on the earth, an abundance of the basic molecular building blocks out of which all known forms of life are constructed. These molecules—amino acids and nucleotides—immersed in scattered shallow pools, would have collided ceaselessly; now and then the collisions would have linked them into the large molecules—proteins, DNA and RNA—which are the essence of the living organism. The linking of small molecules to form large ones would have marked the first step along the path from nonlife to life.

Chemical evolution could have commenced then, in a brief watery period at the beginning of the moon's existence, and been cut off midway in its passage across the threshold, as the water disappeared. If this happened, some of those complex molecules will eventually be found in samples of lunar rock. The detection of such molecules—lying halfway between life and nonlife—would be nearly as important as the discovery of extraterrestrial life itself, because they would reveal one of the ways in which life can appear on a newly created planet.

What is the best place in which to search for these precious molecules? Not on the lunar seas, because we know now, from the analysis of the samples collected in the Apollo landings, that the lunar seas are pools of congealed lava —rocks that once were melted and raised to a temperature of at least 2000 degrees Fahrenheit. At such high temperatures the fragile molecules of life would have been completely disintegrated. If relics of the beginnings of life are to be found anywhere on the moon, they will certainly not be found on the seas.

The lunar highlands, a rough and heavily cratered region adjoining the Sea of Tranquility, are a more promising place in which to search for them. The highlands do not have the look of large, flat lakes of congealed lava. They look like the original surface of the moon just after it was born—a no man's land of jumbled blocks and explosion pits, never melted down since that time. If biologically interesting molecules ever existed on the highlands, they can probably still be found there now. We will have to dig for them, because a fragile molecule lying on the surface would be destroyed in a short time by solar ultraviolet radiation and cosmic rays, but several feet below the surface such molecules may still be preserved.

The highlands are more interesting than the seas for still another reason. Old as the seas may be, the highlands appear to be even older, for photographs clearly show that the materials of each lunar sea fill a basin in the rocky terrain out of which the highlands are formed. Therefore the highlands must have existed before the seas. Moreover, the lunar seas have fewer craters than the highlands, suggesting that they came into existence after the intense bombardment that occurred at the beginning of the moon's life. It is only the highlands that offer lunar explorers the prospect of returning to the very first years of the moon's existence.

The lunar highlands were not at the top of the list of targets for Apollo landings because they are a treacherous, rocky terrain, in which the landing radar of the LM may be confused by multiple echoes, or the craft may come to rest in a dangerously canted position. It would have been unwise to attempt a highlands landing at the beginning of the lunar exploration program, before extensive experience had been acquired in landing techniques.

But a highlands site was the target of Apollo 14, and more highlands landings have been scheduled for later Apollo flights. If the molecular ingredients of life are abundant in the highlands, they will turn up in the samples returned from one of these flights. If such molecules are rare, we may not come across them for years, but the search will go on, because there can be no greater prize in lunar exploration than the discovery of the molecular precursors of evolution.

These results illustrate the scientific value of lunar exploration. The rocks that litter the moon's surface surely contain no living organisms; we know they contain very little gold or silver; but, nonetheless, they are scientifically priceless because of the revelations they can offer regarding the early years of the solar system.

By no means have all scientists been optimistic about the chances of discovering such scientific treasures on the moon. The "hot-moon" school, led by a number of eminent scientists, believes that the moon is similar to the earth, with a molten or partly molten interior, and that its surface has been marked by the same volcanic upheavals and repeated floods of lava that have covered over the original surface of the earth.* Hot-moon scientists point to

* The earth has a core of molten rock and iron at a temperature of 10,000 degrees Fahrenheit and many smaller pockets of molten rock scattered throughout its solid outer layers. When one of these pockets is connected to the surface of the earth by a crack in the solid crust, volcanic upheavals occur, accompanied by great floods of hot, liquid rock. Throughout the earth's history, its surface has been made over by these internal convulsions, which continually bring up new materials that spread out over the surface as lava and congeal to form fresh rock. As a consequence, the average age of rocks on the earth's surface is less than a billion years.

Older rocks are found here and there on the earth, but they are the exception rather than the rule. The oldest rocks—laid down as freshly congealed lava 3.6 billion years ago—are found only in two or three widely scattered locations and are very rare finds. They are not typical of the newness of the earth's surface.

volcanic domes in the Ocean of Storms resembling volcanic islands such as Hawaii. They stress the results of a remote-controlled chemical analysis performed in 1968 by the Surveyor spacecraft, which showed that lunar rocks are made of the same materials, and in roughly the same percentages, as certain rocks on the earth that are known to be of volcanic origin.

Hot-moon scientists would expect most of the rocks on the moon's surface, like those on the surface of the earth, to be less than a billion years old.* They would be surprised to find clues on the moon to the beginning of life or the beginning of the solar system. They argue that the record of the first billion years of the solar system, if it exists on the moon, will be deeply buried and difficult to read.

The "cold-moon" school, led by Harold Urey, holds entirely different views. Cold-moon scientists regard the moon as a very different kind of planet from the earth, containing neither a molten core nor any pockets of molten rock in its interior. They contend that the moon was formed cold, or cooled off shortly after its birth, and that ever since then it has been solid rock straight through to the center. In support of their view they point to the ravaged face of the moon, still bearing the scars of meteorite collisions that must have occurred billions of years ago. If the surface of the moon were continually renewed by fresh lava from the interior, these craters would have long since disappeared.

Cold-moon scientists emphasize the importance of the moon's distorted shape. If the moon were hot, or even lukewarm, the rocks in its interior would be relatively plastic and yielding. Consequently the moon would have assumed a shape very close to that of a perfect sphere, because gravity would pull all parts of the planet's plastic body toward its center. But observations show that the moon's shape is quite far from a perfect sphere. It has an equatorial bulge —a kind of spare tire—about a mile or two in height, and a bulge pointed toward the earth—a sort of moon-nose—also about one mile high.

Cold-moon scientists also point to the significance of the recently discovered mascons—enormous, dense concentrations of mass, of unexplained origin, located in the centers of several of the large lunar seas. These bulges and concentrations of mass could not exist on the surface if the moon's interior were warm and plastic; the interior of the moon would then have yielded under their extra weight, and they would have sunk down into the interior and disappeared.

* That is, less than a billion years ago they rose to the surface as lava and congealed into solid rocks. Of course, the individual atoms in these rocks are far older. They existed as gases in space long before they condensed to form the materials of the solar system.

The fact that they have not done so suggests that the moon's interior is strong, rigid and *cold*.

According to the cold-moon school, the surface of the moon has not been flooded by lava since the earliest years of its existence. The materials on the surface have been worked over by meteorite bombardment but are otherwise the same as they were when first laid down: pieces of primitive moon, with the secrets of the solar system locked inside them. Cold-moon scientists would expect to find many rocks on the moon that were older than the oldest rocks on the earth, and they would even hope to find a few rock fragments that dated back 4.6 billion years to the beginning of the solar system.

Prior to the Apollo 11 landing the evidence was strong on both sides of the dispute, and contradictory. Hot-moon scientists saw many volcanoes—extinct as well as active—in lunar photographs. Cold-moon scientists saw few signs of volcanism in the same photographs. Hot-moon scientists offered proof that the lunar seas—the great dark regions on the moon—are beds of relatively fresh lava, no more than 500 million years old; cold-moon scientists offered proof that the lunar seas are ancient, possibly 4.6 billion years old.

Attitudes hardened as the day of the lunar landing approached, but partisans of both viewpoints believed that Apollo 11 would settle the controversy. A feeling of intense excitement gripped us as we awaited the return of the priceless rocks. Finally, the day came on which the first box of rocks was scheduled to be opened. Tension mounted in the viewing room of the Lunar Receiving Laboratory as scientists and reporters watched the Laboratory staff conduct an elaborate ritual devised by suspicious earthlings to protect themselves and their environment from their first contact with an alien planet. There was only a minute probability that the lunar rocks would contain a lethal microorganism against which life would be defenseless, but the potential danger was great; hence painstaking precautions were taken. Hot-moon and cold-moon scientists were present. We knew that weeks or months would pass before the experiments that would determine the ages of the rocks could be completed, but we also knew that the moment of truth was close at hand.

The results of the age measurements were revealed at a unique conference of lunar scientists in Houston in 1970. One hundred and forty-two Apollo 11 research teams presented their reports to more than a thousand colleagues and reporters who had come to Houston from all parts of the world. Working with small amounts of the precious material, averaging no more than a thimbleful, they had subjected the rocks to every possible variety of investigations, ranging from absurdly simple operations—equivalent to kicking a tire—to complex, delicate laboratory analyses; they had squeezed the rocks under high pressure, heated them to the melting point, examined them under microscopes with polarized light, bombarded them with X rays, and finally taken them apart, atom by atom.

The conference sessions were dreary, as oversized scientific gatherings usually are. Hundreds of people sat in the darkened hall, numbed by exposure to a rapid succession of graphs and charts. The volume of facts and figures paralyzed the mind, yet an undercurrent of excitement gripped us, for we were witnesses to an historic occasion—the first discussion by earthlings of alien materials brought home from another planet.

The highlight of the conference came during the opening session: All the rocks from Tranquility Base turned out to be billions of years old, and some bits of lunar dust were 4.6 billion years old—*as old as the solar system.*

The significance of that single result cannot be exaggerated. It meant that some of the dusty fragments at Tranquility Base had lain on the moon since the earliest years of the solar system, and that through them we could transport ourselves backward in time to the moment when the planets were freshly condensed out of the swirling clouds of the solar nebula. What elements went into the newly created earth? How hot was its surface? How dense was its primitive atmosphere? Were molecules present, of the kind that could lead to the development of life? Not a single fragment of the primitive earth can be found today to answer these questions; but pieces of old moon are available and, in fact, judging by the Apollo 11 samples, abundant.

There was rejoicing in the camp of the cold-moon scientists. They had been prepared to make a long, patient search for those rare objects, the oldest rocks in the solar system, and now they found these precious antiquities strewn about on the moon, waiting to be picked up in the very first hit-and-miss collection. They felt like prospectors who, hoping to find a few bits of gold dust, had stumbled onto a field carpeted with nuggets.

But the confidence of the cold-moon scientists was jolted by another disclosure made at the same meeting. The Apollo 11 experimenters had divided the moon samples into two categories: fine grains of rock and particles of rock dust, pulverized by meteorite bombardment; and large chunks of rocky material that had apparently solidified from molten rock at some time in the moon's past. When the ages of the various samples were measured, the fine grains of rock turned out to be 4.6 billion years old, as we have already noted, but all the large chunks of lava were uniformly 3.6 billion years old—a billion years younger.*

There was a great deal of head-scratching over this peculiar result, and still is today. Why should the finely divided rock particles be of one age, while the solid rocks were of another?

There are nearly as many opinions on the cause of the difference in ages as there are lunar scientists, but there is general agreement on its broad meaning. The fine grains of rock at Tranquility Base must be the broken-up remains of the original lunar surface, formed when the moon was formed 4.6 billion years ago. These rock grains, although fragmented by incessant meteorite bombardment, apparently have not been melted or chemically changed in any way since that time. They are bits of original moon.

The 3.6-billion-year-old rocks, on the other hand, must be the product of a violent event that occurred 3.6 billion years ago. Perhaps the moon was hot enough at that time to produce a major volcanic eruption that spread molten rock over its surface. On the other hand, if the moon was cold then, and solid inside, the rocks could have been melted by some external force, such as the collision of a giant meteorite with the moon's surface.

Whatever the cause of the melting, after the molten layer solidified it was broken up by subsequent collisions with smaller meteorites, and pieces were strewn over a large area. The chunks of lava lying about on the surface of Tranquility Base would then be of these pieces.

Cold-moon scientists insist that the second theory—melting by a meteorite collision—is the only explanation of the melted rock that fits all the evidence. If they are right, and the 3.6-billion-year-old rocks were melted by a meteorite, then these rocks have something important to tell us regarding the earth's history. By a striking coincidence, the oldest rocks on the earth are the same age as the rocks found at Tranquility Base. Apparently they, too, were melted 3.6 billion years ago. Is it possible that the same catastrophe overtook the earth and the moon at that time? If the moon ran into a hail of giant meteorites that melted large parts of it, then the earth must have been even more heavily bombarded, because its strong gravitational force would pull these meteorites toward it faster and they would hit the earth at an even greater speed than they hit the moon. It is likely that all life would have been wiped out on the earth

* With one exception, an aberrant fragment 4.4 billion years old, nicknamed the Luny Rock by Apollo 11 experimenters.

by the heavy bombardment, only to rise again, phoenix-like, out of the original molecular ingredient, when the surface of our planet had cooled sufficiently.

Two geneses on one planet? That would indeed be a remarkable tribute to the strength of the evolutionary process. Moreover, since the oldest fossils found on the earth also go back nearly 3.6 billion years, the second genesis, if it happened, would force us to a drastic revision of our estimate on the length of time required for the evolution of life out of nonliving chemicals.

These ideas, suggested by the results of the first lunar landings, illustrate how the earth's history can be illuminated by the study of the moon. Later landings may invalidate the conclusions from the first lunar flights, or lead to even more interesting possibilities. Whatever is in store for us in the future exploration of the moon, the Apollo samples have already yielded the most extraordinary result that could have been hoped for. We are now certain that some of the materials on the moon have lain there for 4.6 billion years, since the time when the moon and the earth were freshly condensed out of the parent cloud of the solar system. There is no longer any doubt that the record of the past, missing on the earth, can be deciphered on the airless, waterless moon. The moon is the Rosetta Stone of the planets.

THE ANCIENT SURFACE OF THE MOON. The pitted landscape of the moon, carrying the scars of tens of thousands of meteorite collisions, bears witness to the excellent state of preservation of the lunar surface. The largest craters are visible in the photograph at left, made by joining two separate pictures of the half moon. The area outlined in white is shown in greater detail in the photograph at right, taken with the 100-inch telescope on Mount Wilson.

Meteorites have produced similar craters on the earth throughout its history, but most of these have been erased by the wearing action of wind and running water. Only the most recently formed craters, such as the Arizona Meteorite Crater (*below*), are still visible. The Arizona Crater, approximately one mile in diameter, is very similar in shape to craters of the same size on the moon. It was formed in the recent past, about 30,000 years ago, and will have vanished in 10 million years. Similar craters on the moon last for billions of years because the moon has no water or air and very little erosion.

A STILL-CLOSER VIEW OF THE MOON: LUNAR CRATERS, OLD AND NEW. The Ranger 9 photograph at right shows an area of 5 square miles on the moon's surface. Many small craters with sharply defined edges are visible. These craters were formed relatively recently in the history of the moon.

The photograph also reveals other craters of the same size, which are partly or almost entirely filled in and have rounded edges. These partly filled craters were formed earlier in the history of the moon. Their edges have been worn away by the constant bombardment of small meteorites over a period of a billion years.

Measurements on the photographs show that about 50 feet of lunar surface material has been redistributed by meteorite bombardment during the history of the moon. This rate of erosion is 10,000 times smaller than the erosion produced on the earth by wind and running water. The negligible rate of erosion supports the view that the moon's surface changes very slowly.

The area outlined in the white rectangle is shown in detail in the inset. The crater at lower left in the inset is 150 feet wide and 30 feet deep. This photograph was taken 1½ seconds before the Ranger spacecraft crashed into the surface of the moon. The arrow marks the impact point.

The second phase of lunar exploration commenced with the flights of the Surveyor spacecraft. Surveyor, unlike Ranger, carried retrorockets designed to slow its descent to the moon. Cameras and instruments were dropped gently onto the surface and began to operate under radio control. Later Surveyors carried instruments to analyze the chemicals in the lunar soil and the strength and texture of the soil in preparation for the manned landing.

A Surveyor camera photographed the 15-inch rock shown below, lying a few feet from the spacecraft and partly buried in a fine debris of pulverized material thrown out in meteorite collisions over millions of years.

THE FIRST LANDING. Eagle touched down on the surface of the Sea of Tranquility on July 20, 1969, at 4:18 p.m. EDT. The photographs at left, taken by Armstrong, show Aldrin emerging from the LM thirty minutes after the landing. He appears to float down the ladder because the force of gravity on the moon is one-sixth of the earth's gravity.

The reduced force of lunar gravity also explains the surprising tilts of the astronauts' bodies as they loped over the surface. Because they weighed so little, the friction between their boots and the ground was diminished. To gain the extra friction needed for a change in course, they had to lean into the direction they wished to follow and push against the ground at an exaggerated angle. Aldrin reported later, "Like a football player, you just have to split out to the side and cut a little bit."

"THE SURFACE IS FINE AND POWDERY," Armstrong said. "I can pick it up loosely with my toe. It does adhere in fine layers like powdered charcoal to the sole and sides of my boots. I only go in a small fraction of an inch. Maybe an inch, but I can see the footprints of my boots and the treads in the fine sandy particles." Armstrong's boots sank in six or seven inches in some places. The Armstrong footprint in the photograph below will remain clearly visible for a million years.

MOON DUST. One of the surprises of the lunar landing resulted from the examination of the moon dust under a miscroscope. A substantial part of the dust particles consisted of tiny glass beads, apparently formed when meteorite impacts melted part of the moon's surface and threw out a fine spray of molten droplets. Being very small—a few thousandths of an inch in diameter—the droplets cooled rapidly to form glassy spheres, rather than irregular chunks of lava-like material. The photograph of a moon-dust sample (*below*), enlarged 100 times, shows several glass beads mixed with irregular fragments of lunar rock.

The presence of the glass beads explains Armstrong's comment, radioed back from Tranquility Base, "These rocks are rather slippery." Although he did not know it, he was skating on a surface of ball bearings.

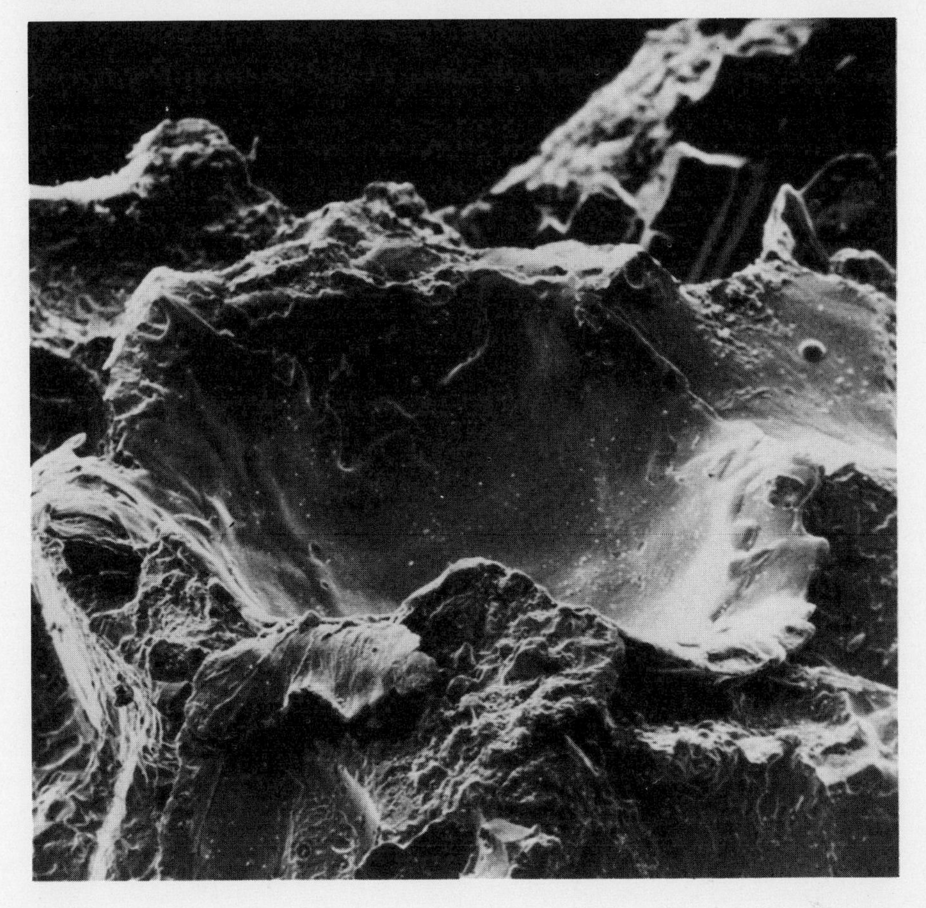

MINI-CRATERS. Most of the Tranquility Base rocks were scarred by small pits, between one-hundredth and one-tenth of an inch in diameter. The photograph (*above*) shows one of these pits enlarged 1000 times. The pits are mini-craters blasted out by the impact of tiny meteorits, smaller than a pinhead. These minute particles hit the moon's surface at speeds up to 100,000 miles per hour, releasing more energy per pound than a TNT explosion. Billions of particles of this size also penetrate into the earth's atmosphere each day but are burned up by friction before they reach the surface. As they disintegrate in the atmosphere, they leave behind the incandescent rails of vaporized rock known as "falling stars."

A TYPICAL SCENE ON THE LUNAR SEAS. Most of the surface of the moon resembles this monotonous terrain, studded with small craters and littered with rock fragments. The powdery soil is heavily trampled by the two-hour exploration. The landscapes lack color, being variously reported by the astronauts as a uniform dark gray, or sometimes gray with a hint of a warmer tone. But the astronauts were fascinated by the harsh contrasts of light and shadow, unsoftened by an atmosphere. Armstrong reported, "It has a stark beauty all its own. It's like much of the high desert of the United States."

The photograph shows Aldrin setting up the solar wind experiment, in which a sheet of aluminum foil is placed to catch fast-moving particles blown off the surface of the sun. Because of their high speed, the particles penetrate the outer layers of the foil and are trapped within it. The foil is rolled up like a window shade, carried back to earth, and heated in the laboratory to dislodge the trapped particles. The experiment yielded direct information on the chemical elements in the body of the sun.

EVIDENCE FOR A HOT MOON. Signs of volcanism in lunar photographs apparently were confirmed by Apollo 11 results, showing that lunar rocks resemble lava on the earth in their porous structure and chemical composition. The photograph below shows one of the porous, lava-like lunar rocks collected from Tranquility Base.

The evidence for lunar volcanism supports the "hot-moon" scientists, who argue that the moon's interior was molten at some time in the past and has been the scene of extensive volcanic eruptions that have covered its surface repeatedly with floods of fresh lava, obliterating the record of the moon's past.

LUNAR VOLCANISM is also suggested by the Hyginus Rille, a huge crack in the face of the moon (*right*). This rille is about 2 miles wide and 100 miles long. The area outlined in white is shown (*below, right*) in an overhead view, revealing a row of craters inside the rille spaced along much of its length. These cannot be the result of meteorite impacts, which would produce a random pattern of craters over the area. The Hyginus Rille craters must be volcanic, the entire rille possibly being the surface manifestation of a deep-seated rupture within the moon's body.

EVIDENCE FOR A COLD MOON. The Apollo 12 astronauts set up a scientific station which functioned under remote control for more than a year. The lunar station included a seismometer for measuring moonquake activity, which revealed that quakes occur far less often than on the earth, indicating that most of the moon is cold and stable at the present time.

Additional evidence for the cold-moon theory came from a second instrument, which measured the moon's magnetism. Changes in magnetism were observed when blasts of particles from the sun swept past the moon. These changes depend on the electrical resistance of lunar rocks, which depends, in turn, on the temperature within the moon. The data indicated a high electrical resistance, and therefore a cold moon. The conflict with evidence for the hot moon has not yet been resolved. One explanation is that the moon was hot for a brief period in its youth, then cooled rapidly and has been geologically lifeless ever since.

THE LUNAR HIGHLANDS, a rough and heavily cratered region (*above*), are older than the lunar seas. They look like the original surface of the moon just after it was born—a no man's land of jumbled blocks and explosion pits, never melted down since the beginning of time in the solar system. The highlands, objectives of the Apollo 15 and 16 landings, were not among the first Apollo landing sites because they are a hazardous terrain.

The photograph of the "shoreline" of the Ocean of Storms (*right*) shows one reason why the highlands are considered to be older. The smooth surface of the Ocean of Storms laps against the adjoining highlands terrain and seems to fill a basin in the rocks out of which the highlands were formed. Presumably the highlands existed first, and the basin was blasted out later by the impact of a giant meteorite. Subsequently lava accumulated in the basin, just as the waters of the earth accumulated in the natural basins of the earth's crust. This conclusion is strengthened by the fact that the highlands have a greater density of craters than the seas, indicating they have been bombarded by meteorites for a longer time.

☼

8 Venus, Mars and Jupiter

NINE planets circle the sun. Six of these—Mercury, Venus, Earth, Mars, Jupiter and Saturn—were known to the ancients, while three—Uranus, Neptune and Pluto—were discovered in modern times. Mercury is the closest of the nine to the sun. It is a small planet, less than half the size of the earth, and scarcely larger than the moon. Its rocky, barren surface, alternately baked on the side facing the sun and frozen on the side facing away, is extremely inhospitable to life. The planet is difficult to reach by rocket from the earth, because of its closeness to the sun, and it is unlikely that we will learn more about it than we know for many years to come.

Moving outward from the sun beyond Mercury, we come to Venus. Venus is the earth's closest planetary neighbor. It is also our sister planet, closely similar to the earth in size and weight, and situated at a distance from the sun which is not very different. The surface of Venus is completely covered by clouds, and conditions on the planet have always been an enigma, yet romantic hope has flourished that beneath these clouds lie teeming masses of flora and fauna. In 1686 de Fontenelle, in his book *Conversation on the Plurality of Worlds,* described the characteristics he expected to find in the people on Venus:

> —I can tell from here . . . what the inhabitants of Venus are like; they resemble the Moors of Granada; a small black people, burned by the sun, full of wit and fire, always in love, writing verse, fond of music, arranging festivals, dances and tournaments every day.

In fact, Venus should provide an even more agreeable climate for living organisms than the earth. The planet is 70 million miles from the sun, while the distance of the earth is 93 million miles. Because Venus is closer it receives twice the intensity of the sunlight falling on the earth; and although its heavy cloud cover keeps out some of this solar energy, we can still estimate that on Venus the average temperature at the latitude of London should be a comfortable 80 degrees Fahrenheit, or approximately the same as the balmy temperatures of the islands of the Caribbean.

But twelve years ago it was discovered that this is not at all the case. The climate of Venus, far from being balmy, appears to be very uncomfortable and of such a character as to discourage any possible hope of finding life on the surface of the planet. From measurements of the intensity of the radiation emitted by Venus it has been deduced that the temperature of the surface is a sizzling 800 degrees Fahrenheit, which is well above the melting point of lead. It is certain that no organism remotely resembling terrestrial life could survive in such heat.

Yet hope lingered on for the discovery of a green world on Venus. Some astronomers argued that the intense radiation might come from the atmosphere of Venus and not from its surface. Others suggested that life might be supported at the north and south poles, which should be cooler.

In 1967 Russian and American spacecraft reached the planet and carried out measurements that removed the last trace of doubt regarding the high temperature of its surface. Launched within two days of each other in June, 1967, the U.S.S.R. spacecraft Venus 4 and the U.S. spacecraft Mariner 5 traversed elliptical paths of approximately 200 million miles and arrived in the vicinity of the planet four months later. The histories of the spacecraft diverged as they drew near Venus. Drawn toward the planet by its gravitational pull, Mariner 5 swerved across the dark side, over a portion of the sunlit hemisphere, and out into space. Venus 4, heading directly for the planet on a collision course, ejected a capsule which parachuted downward near the equator, radioing back information as it descended, in an extraordinary feat of planetary exploration.

The two spacecraft measured the temperature in different ways, but their data led to the same conclusion. Two subsequent U.S.S.R. flights in 1969 by similar spacecraft—Venera 5 and 6—provided further confirmation. Venus is indeed hot enough to melt lead, and there is no reasonable chance of finding life on its surface.

Why is Venus so hot? The U.S.S.R. and U.S. spacecraft carried out other measurements which revealed the answer to this question. According to information radioed back to the earth from the spacecraft, the atmosphere of Venus consists primarily of a heavy layer of carbon dioxide, about 10,000 times more than is in the atmosphere of the earth. The dense atmosphere of carbon dioxide acts as an insulating blanket which seals in the planet's heat and prevents it from escaping to space. The trapped heat raises the surface to a far higher temperature than it would have otherwise. Calculations based on the insulating properties of carbon dioxide show that the temperature of Venus easily could be raised to 800 degrees as a result of this effect.*

* The temperature of the earth would also be unbearable if the planet were covered by a blanket of carbon dioxide as dense as that on Venus.

In addition to its instruments for measuring temperature, the Venus 4 spacecraft also carried instruments designed to detect water. These indicated the presence of a moderate amount of water vapor in the atmosphere—sufficient, if condensed to a liquid, to cover the surface of Venus to a depth of one foot.

It is a great puzzle to students of the planets that no more water than this was found. According to current views on the origin of the planets, Venus and the earth condensed out of the same materials, containing similar amounts of water. During the condensation water was trapped in the interior of the earth. Later the trapped water rose to the surface and escaped through cracks in the crust to fill the oceans. The water in the earth's oceans, if spread uniformly over the surface of the planet, would form a layer about 8000 feet deep. A layer of water of approximately the same thickness should also exist on Venus. Because of the high temperature on Venus' surface, the water would be present not in liquid form but in the form of water vapor in the atmosphere. Venus 4 showed that most of this water is missing.

Thus, by terrestrial standards Venus is a dry, hot planet and not a very suitable environment for the development of life.

Why did two planets, probably formed out of similar materials and situated at comparable distances from the sun, evolve along different paths? Why is the surface of Venus baked by a searing heat, while the earth luxuriates in a climate friendly to all known forms of life?

The high temperature on Venus is explained by the abundance of carbon dioxide in its atmosphere, but this gas should be equally abundant in the atmosphere of the earth. When we understand why there is less carbon dioxide on the earth than on Venus, we will also understand their differences in climate. To some extent, the explanation must be connected with the presence of life on the earth. At the present time, much of the carbon dioxide in the earth's atmosphere is removed by marine animals, which absorb this gas in sea water and convert it within their bodies to solid substances known as carbonates. Sea shells, for example, are nearly pure calcium carbonate. Today the upper layers of the earth's crust contain a thick layer of carbonate, formed by the compressed shells of countless mollusks and crustaceans that died long ago; and locked up in these carbonates is a large part of the carbon dioxide that would otherwise blight our atmosphere, as it blights the atmosphere of our less fortunate neighbor planet.

When the earth was young, life was either absent or scarce, and carbon dioxide could not have been removed in this way. But the gas still could have been absorbed from the atmosphere of the young earth by other chemical changes, not involving living organisms. In these reactions, atmospheric carbon dioxide combines with rocks to form carbonates, somewhat as oxygen in the atmosphere combines with iron to form rust. However, such reactions do not take place at an appreciable rate if the rocks on the surface of the planet are very hot and dry. Therefore they cannot occur on Venus.

In other words, carbon dioxide is removed from the atmosphere of the earth in many ways but cannot be removed in any way from the atmosphere of Venus. It is understandable that Venus now has a great abundance of this gas in its atmosphere.

The present concentration of carbon dioxide on Venus must have built up slowly in the atmosphere over hundreds of millions of years. Why did life not develop in this initial period, before too much of the harmful, heat-insulating gas accumulated? If life had once spread over Venus, its ability to absorb carbon dioxide could have kept the planet comfortable forever after. The answer must be connected with the fact that Venus is closer to the sun and was somewhat warmer than the earth at the start. Because of the moderately higher temperatures on the surface, the rocks on Venus did not absorb the slowly accumulating carbon dioxide quite as rapidly as rocks on the earth. The concentration of the gas built up, retaining the planet's heat and making its surface still hotter, until finally the conditions needed for the development of life were permanently destroyed. With no life present to absorb carbon dioxide, the gas continued to accumulate in the atmosphere unchecked. Thus the chain of events commenced that led to the oven-like conditions of today.

How far from the sun must a planet be to maintain a comfortable temperature? As yet, we do not know; the spacecraft results tell us only that Venus was too close, and that was its undoing. If it had been only a few million miles farther away, the temperature of its surface might have climbed slowly enough to permit life to gain a toehold; and once life began, it would have held the abundance of carbon dioxide in check, and prevented the temperature from climbing out of bounds subsequently. Having lost its chance to harbor life when the solar system was first formed, Venus could never again recapture the opportunity.

Beyond Venus and beyond the earth lies the planet Mars. Mars circles the sun at a distance of 140 million miles, one and one-half times more distant than the earth. The density of Mars, like that of Venus, is about the same as the density of the rocks lying on the surface of the earth; for this reason Mars is believed to be composed of rocky materials similar to those on our planet. The atmosphere of Mars is rather thin, perhaps a hundredth as dense as the earth's atmosphere. On Mars, unlike Venus, only a trace of clouds exists, and the features of the Martian surface are not appreciably obscured by them. Dust storms and haze occasionally are conspicuous, but most of the time the face of the planet is open to photographic surveillance.

In spite of the thinness of the Martian air, it is impossible to obtain good photographs of Mars from telescopes on the earth because of the blurring effect of the earth's atmosphere on rays of light reaching us from it. No features of the surface of Mars can be seen from the earth, no matter how large the telescope in which the planet is viewed, unless they are 50 miles or more in diameter. It is impossible to tell from the earth whether Mars has mountains, ocean beds, or any features that might indicate the presence of life.

Far better pictures of Mars were taken by the Mariner 6 and 7 spacecraft in 1969, as they swept past the planet at a distance of a few thousand miles. The clearest of the Mariner photographs revealed features as small as 100 feet across. They showed that the surface of Mars is marked by a large number of craters, presumably produced by meteorite collisions. There are relatively fewer craters than on the moon, but far more than on the earth.

The Mariner photographs suggest that Mars stands midway between the moon and the earth as a planetary body, its surface being older and better preserved than the surface of the earth but not as well preserved as the surface of the moon. The factors that make Mars more promising than the moon as an abode for life and for man—its thin but appreciable atmosphere and trace of moisture—have diminished its value as a source of clues to the early history of the solar system.

Will life be found on Mars? Possibly. Laboratory experiments have shown that plants can exist in the dry, cold, oxygen-poor Martian climate. In these experiments, some plants exposed to a simulated Martian climate died after a month, but new shoots grew up in their places. Other plants were injured but survived.

Since earth plants barely survive in a Martian environment but do not flourish there, it seems that the Martian plant population, if it exists at all, must be stunted and meager. However, we must allow for the fact that if there is life on Mars, it must have evolved during a time when there was an abundance of water on the surface.* Living organisms can evolve out of nonliving molecular

* No evidence exists to exclude the probability that Mars has seen wetter days in its past. The Mariner photographs taken in 1969 do not show sufficient detail to reveal subtle traces of an earlier period of extensive erosion by rain and running water. Photographs showing more detail may be obtained in 1971 by cameras on an improved version of the Mariner spacecraft. Until such photographs are available, this fascinating question remains open.

ingredients only if those ingredients are dissolved in an ample supply of water, in which they can move freely and collide with one another again and again. Repeated collisions between neighboring molecules are essential for the assembly of the large molecules of life—proteins and DNA—out of smaller ones. Mars may be lifeless; but if life exists there it is likely to be descended from a golden age on the planet when its climate rivaled that on the earth.

If the transition to the dry climate of today occurred slowly enough, over a period of millions of years and a like number of generations, Martian life could have adapted progressively to the gradual onset of severe conditions. During this long period, the weakest individuals in each generation would be eliminated and the hardiest would remain, propagating their qualities of strength to their descendants. There seems no reason to doubt that varied and interesting forms could exist on Mars today as a result of this long-continued process of natural selection, *if* the planet once had water.

Martian organisms, highly specialized for survival on a nearly water-free and airless planet, doubtless would present an unusual appearance; their forms, internal arrangements and methods of reproduction might seem bizarre; the fundamental differences between plants and animals, as we know them, might be blurred. Nonetheless, this extraterrestrial life would have much to teach us about the nature of life on the earth, for the basic chemistry of Martian life—product of an independent line of evolution, and adapted to markedly different conditions—probably would not be identical with the chemistry of terrestrial life. From the comparison between the two living structures, parallel but distinct, we would gain insights into the metabolism of all living organisms, including man, that we could not acquire in decades of laboratory research on earth. In the realms of medicine and biology this prospect stands out as the greatest potential contribution of planetary exploration.

There is a larger significance to the search for life on Mars. The sun is one of 100 billion stars belonging to the cluster we call our Galaxy. According to the best evidence, many if not most of these stars are circled by families of planets. Ten billion other galaxies, each with one hundred billion stars—and probably planets—are within the range of large telescopes. Perhaps only a small fraction of these are earthlike planets, but that could mean millions of earthlike planets in our Galaxy alone. It may be that all earthlike planets, except the earth itself, are barren bodies of rock. However, if life appeared spontaneously on the earth, it could appear elsewhere. What is the probability of this happening? If it is as low as one in a billion, we are alone in this corner of the universe. If it is as high as one in ten, or one in one hundred, inhabited planets must be everywhere in the galaxy, and the most extraordinary experiences of *Homo sapiens* still lie ahead of him.

Mars offers the best hope of learning the answer to this question. It is one of only two earthlike planets in the solar system that could conceivably support life. We know that on one of these—the earth—life has in fact developed. That may be a unique accident, but it is unlikely that two such accidents would have occurred in one solar system. If life—or the remains of life—are discovered on Mars, we will be forced to conclude that the development of life out of nonlife is not a rare accident but a relatively probable event. No scientific discovery more significant for mankind can be imagined.

The search for life on Mars has already begun. Living organisms betray their existence by chemical changes they produce in their environment, which can be detected by remote-controlled instruments. In 1969 two Mariner spacecraft flew past the planet at a distance of a few thousand miles with instruments capable of sensing some of the chemicals associated with life. One of these instruments measured infrared radiation in a region of wavelengths where the gas methane has characteristic absorption bands. Methane is released by decaying vegetation, but, being relatively unstable, it does not last very long in the atmosphere unless plants are present to continually renew the supply.

Methane was not detected, a negative result that seems to quench our hopes of finding Martian life. However, this conclusion would be premature, for the instruments, never closer to the surface of Mars than 1800 miles, could not detect these gases if they were present in extremely small amounts. The limit on the sensitivity of the instruments corresponded to a concentration of one part per million of methane, which is about the same as the concentration of methane in the earth's atmosphere. Thus, a Mars flora could exist and be nearly as abundant as the vegetation on the surface of the earth, and still have escaped detection in this experiment.

Another Mariner instrument, designed for the detection of nitrogen, provided a second test for life on Mars. Nitrogen, like methane, is a product of the cycle of growth and decay in living organisms. The instrument failed to indicate the presence of the critical gas. However, the smallest amount of nitrogen that the Mariner instruments could detect is approximately equal to the amount of nitrogen *of biological origin* in the earth's atmosphere.* As in the case of the methane experiment, this experiment does not exclude the pres-

* Only a small part of the earth's atmospheric nitrogen is associated with life; 99.99 percent came from the interior of the planet in volcanic gases, and would be present in the atmosphere even if the earth were lifeless.

ence of life on Mars. It only indicates that life cannot be as abundant as, or more abundant than, life on the earth.

A more sensitive test for life on Mars is planned for 1975. In that year an improved version of the Mariner spacecraft will be launched, containing a package of instruments to be dropped on the surface of Mars as the main spacecraft flies by. According to one design, the landing craft will contain a long sticky string that will be thrown out across the landing site and then reeled back into the body of the craft, presumably with particles of Martian soil adhering to it. The particles will be deposited in culture dishes containing ingredients that make excellent food for terrestrial bacteria. This food is made up of atoms of carbon, nitrogen, oxygen and other elements. If microorganisms are present on Mars and their chemistry resembles that of bacteria on the earth, they will consume the food and multiply. The heart of the experiment lies in the fact that the carbon in the food is *radioactive* carbon, and not the nonradioactive carbon normally found in nature. The radioactive carbon, if ingested by the Martian organisms, would be incorporated into the chemicals of their cells in the reactions that constitute the life processes of the organisms. In these reactions, some of the radioactive carbon would be combined with oxygen to make radioactive carbon dioxide. The radioactive carbon dioxide would be exhaled by the bacteria.

Adjoining the chamber with the culture dishes would be a second chamber containing an instrument sensitive to radioactive substances. The second chamber is separated from the first by a filter through which no particle larger than a molecule of gas can penetrate. The food particles containing the original radioactive carbon cannot pass between the two chambers, but carbon dioxide gas can do so. The instrument will send a signal to the earth if it detects radioactivity. The receipt of this signal would indicate that a Martian microorganism has been eating the food containing radioactive carbon.

The experiment is ingenious, but it will work only if Martian life is similar to earth life in its basic chemistry. No one has been able to invest a plausible kind of life chemistry that is completely different from the carbon-based chemistry of earth life, but the failure probably reflects the limited imagination of the scientist, rather than a true limit on the number of chemical possibilities for life. For this reason, life on Mars may remain undetected until a manned expedition reaches the planet.

Beyond Mars there is a large gap in the distribution of the planets. We might expect to find a planetary body located outside the orbit of Mars, about three times the earth's distance from the sun; but instead we find only a large number of small bodies—planetesimals—circling in a ring. These are called asteroids. Occasionally, collisions between these bodies, or perhaps the gravitational pull of Jupiter, the next planet past Mars, will pull one of them out of its orbit and into a collision course with the earth. It is believed that many, if not all, of the meteorites which hit the earth have this origin. Examination of the meteorites that survive the searing passage through the earth's atmosphere reveals them to be pieces of rock and iron with a rather complex physical and chemical history. Many of these meteorites appear to have been pulverized at some point in their early history and compacted again into their present form. All this evidence together suggests that there may once have been a group of planetesimals of substantial size in orbit between Mars and Jupiter. For some reason, these planetesimals did not reach the ultimate stage of accumulation into a planetary body, as did the other objects in the solar system; or, if they did, they were disintegrated again in a subsequent catastrophe.

Five planets lie outside the orbit of the asteriods. These are the giant planets —Jupiter, Saturn, Uranus and Neptune—and the small planet Pluto.

Pluto, found in 1930, was the ninth and last planet to be discovered in the solar system. Its orbit is farther from the sun than that of any other planet and probably marks the outer boundary of the solar system. Because Pluto is so far away, we have been able to learn very little about it, except that it appears to be a body similar in size and composition to the earth. It must be a frozen, silent world, far too cold to support any form of life.

More is known about the giant planets. They are approximately 10 times larger than the earth and 100 times more massive, but considerably lower in density. In general, their density is about the same as that of water; Saturn, in fact, is less dense than water; it would float in the bathtub if you could get it in.

The giant planets are less dense than the earth and its neighbors because they are made up largely of the lightest elements, hydrogen and helium. These elements make up most of the matter in the universe; they also constitute most of the matter in the sun and in the giant planets; but for some reason not clearly understood, they are missing from the earth and inner planets. Perhaps the particles and radiation emitted from the sun in its early years blasted the hydrogen and helium out of the inner parts of the solar system, from which the earth was formed, while the outer regions, out of which the giant planets were formed, were too far away to be affected by the blasting action.

Jupiter is the largest of the giant planets and the most massive planet in the solar system. It is 11 times the size of the earth and 318 times as heavy. On a planet as large as Jupiter, the force of gravity is so great* that most of the gases of the planet's original atmosphere will remain with it throughout its lifetime. Not even the lightest gases, hydrogen and helium, can escape. The atmosphere of Jupiter contains these gases in abundance, and it also contains the gases that are common compounds of hydrogen. These compounds—ammonia, methane and water vapor—were present in abundance in the primitive atmosphere of the earth and are believed to have played a critical role in the events that led to the development of life on our planet. Their importance in evolution on the earth has ended, and they have long since escaped, but their continued presence on Jupiter leads us to wonder whether at least the initial steps along the path to life have not also occurred on that planet. At first this seems unlikely, because Jupiter is nearly 500 million miles from the sun and 5 times as distant as the earth, and receives very little solar heat. Its temperature should be quite low, probably too low to support the chemical reactions necessary for life.

Yet there is a ray of hope in the situation. The measured temperature of Jupiter is a frigid —300 degrees Fahrenheit, but this is the temperature at the level of the clouds covering the surface of the planet. Conditions beneath the clouds are concealed from our view. But below the clouds, and closer to the surface of the planet, the temperature must be considerably higher. This is true on the earth, where the temperature of the air near the ground has been heated by the absorption of energy from the sun. For example, the temperature at the height of a jet flying above the clouds at 30,000 feet is usually about minus 60 degrees Fahrenheit. We do not know whether Jupiter has a well-defined surface or, perhaps, an atmosphere that grows steadily denser until it turns imperceptibly to a liquid; but, with or without a surface, Jupiter probably has a region, at some depth below its clouds, in which the temperature passes through a comfortable range for the development and continued support of life. In this region exist the gases out of which life is believed to have evolved on the earth in its early years. Perhaps a kind of life has developed on Jupiter as well. It could not be the oxygen-breathing life with which we are familiar; we assume that it would be quite different; we will know only when—perhaps at the turn of the century—elaborately instrumented spacecraft make their maiden trip to the giant planet and complete the long voyage home.

* An average-sized man would weigh 400 pounds on Jupiter.

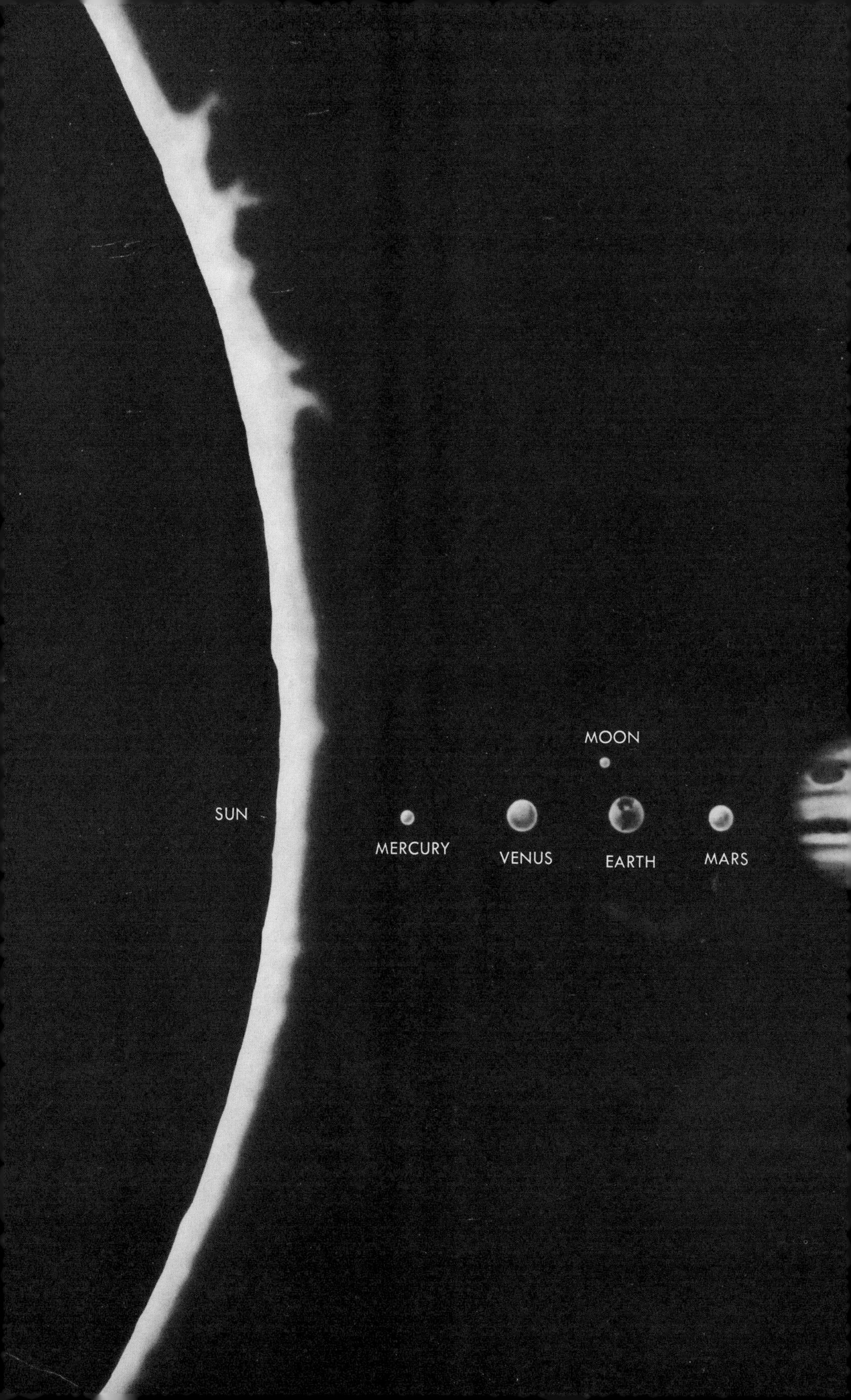
MOON
SUN
MERCURY
VENUS
EARTH
MARS

THE SUN AND THE PLANETS: RELATIVE SIZES. The sun and planets, arranged in order of their distance from the center of the solar system, are shown here in proportion to their actual sizes. The sun is one million miles in diameter, or 13 times the size of the largest planet, Jupiter. Only a small part of the sun's rim can be displayed on the scale of this drawing.

The inner planets—Mercury, Venus, Earth and Mars—are called the terrestrial planets because they are believed to be composed of the same mixture of rocky materials and nickel-iron which make up the body of the earth. The moon, which is only slightly smaller than the planet Mercury, is often grouped with the terrestrial planets.

Jupiter, Saturn, Uranus and Neptune are called the giant planets because they are roughly 100 to 300 times more massive and 5 to 10 times larger in diameter than the terrestrial planets. The giant planets are made up principally of the lightest elements—hydrogen and helium—with only a small percentage of the iron and rocklike materials which compose the bulk of the earth and the other terrestrial planets.

Pluto, the outermost planet, is probably similar in size and composition to the terrestrial planets. It is a frozen and surely lifeless world, circling the sun at a distance of 4 billion miles.

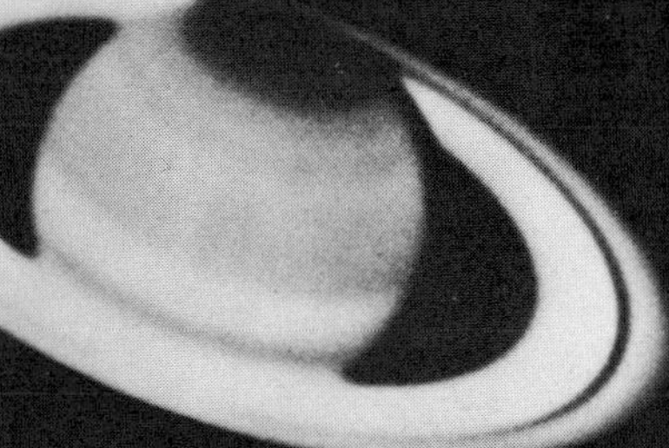

SATURN

URANUS NEPTUNE PLUTO

1
3

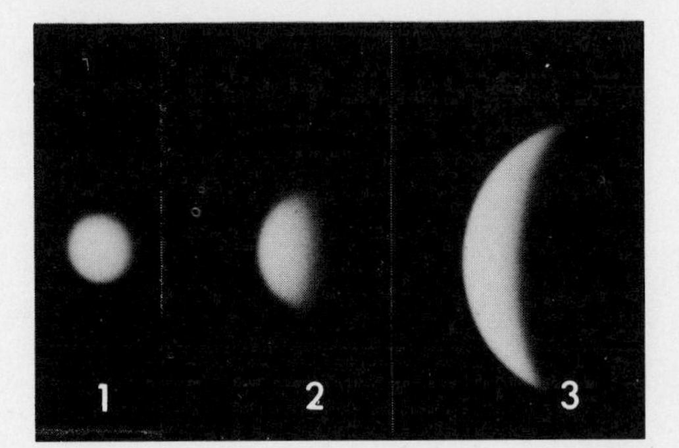

VENUS: SISTER PLANET OF THE EARTH. Venus nearly equals the earth in size and weight. The planet circles the sun inside the earth's orbit (*left*), completing one circuit in 226 earth days. Until recently it was thought to offer a balmy and pleasant climate for life.

As Venus revolves around the sun it goes through phases similar to those of the moon. The phases of Venus are visible in the photograph above, taken with the 36-inch telescope at Lowell Observatory. At the "full Venus" (1) the planet is on the opposite side of the solar system from the earth and its face is fully illuminated by the sun. At the "half Venus" (2), the planet has moved halfway around its orbit toward the earth. The "new Venus" (3) is on the same side of the sun as the earth; because it is directly between the earth and the sun, it can hardly be seen, although at this point in its orbit, it is at minimum distance and assumes its largest apparent size.

A heavy cover of clouds veils Venus at all times. Breaks may occur in the clouds but would not be visible in our telescopes because of the blurring effect of the earth's atmosphere, which obscures all features less than 50 miles across. The detailed appearance of Venus will remain a mystery until the planet is photographed from close range by a passing spacecraft in 1973.

EXPLORATION OF VENUS BY SPACECRAFT. The U.S.S.R. spacecraft Venus 4 (*below*) and the U.S. spacecraft Mariner 5 (*right*) were launched toward Venus within two days of each other in June, 1967, and arrived in the vicinity of the planet four months later after journeys of 217 million miles. Mariner 5 whipped around the planet and continued on around the sun. Just before Mariner 5 vanished behind Venus, and again just after it emerged from the shadow of the planet, radio signals from the spacecraft probed the Venus atmosphere on their way to the earth. Scientists on the earth were able to determine in this way the atmospheric pressure and temperature near the Venus surface.

Venus 4 headed directly for the dark side of the planet and crashed near the equator. As Venus 4 entered the atmosphere, a daughter probe separated from the main spacecraft and parachuted to a soft landing, radioing information on atmospheric conditions during its descent. The photograph below shows the Venus 4 landing craft somewhere in the U.S.S.R., after a test of its parachutes.

Construction details of the U.S.S.R. spacecraft are not known. The Mariner spacecraft is shown at right. Scientific instruments are housed in eight compartments making up the body of the spacecraft (1). A dish-shaped antenna (2) beams the main radio signal to the earth. The dish swivels about in space and can be pointed at the earth by radio command. The four paddles (3) which give the spacecraft its windmill appearance are solar panels covered with photoelectric cells, which generate 550 watts of electricity from sunlight. The length of the spacecraft is 8 feet.

Both the U.S. and U.S.S.R. results confirmed that the temperature on the surface of Venus is 800 degrees Fahrenheit. Lead melts at this temperature, and all forms of life conceivable to us are impossible. The spacecraft results also indicated that the atmosphere consists mainly of carbon dioxide, which is responsible for the ovenlike condition of Venus. It seals in the planet's ground heat and prevents it from escaping to space. The earth had a similar amount of carbon dioxide, but the gas has been absorbed from our atmosphere by chemical reactions with rocks and living organisms. If the earth had been a few million miles closer to the sun, and therefore a little warmer at the start of its existence, these reactions would have been less effective, and our planet, like Venus, might have become a lifeless inferno. The Venus experiments suggest that earthlike planets with a comfortable climate for life may be rarer than previously believed.

3
2
1

MARS PHOTOGRAPHED FROM EARTH: THE CHANGE OF SEASONS. Photographs of Mars from earth (*right*) reveal striking changes during the Martian year, resembling the change of seasons on the earth. The photographs were taken four months apart, at times corresponding to the fall (*above, right*) and winter (*below, right*), in the Martian southern hemisphere. The most conspicuous feature of these photographs is the "polar cap," resembling the cover of ice and snow at the poles of the earth. The polar cap grows in size from 200 miles in the upper photograph to a maximum diameter of 2000 miles in the lower photograph. As a result of Mariner measurements, the polar caps are now generally believed to be mainly composed of frozen carbon dioxide, that is, dry ice, rather than frozen moisture.

Parts of the surface of the planet display seasonal changes in color, ranging from a light yellow-brown to a dark blue-green, which suggest the blooming of a cover of vegetation during the Martian spring. The dark markings fade away during the Martian summer. These variations are clearly visible at right. The significance of the seasonal changes is disputed; some observers believe they prove the existence of plant life on Mars, while others assert that they indicate a lifeless cycle of chemical reactions.

CANALS ON MARS. In 1877, the Italian astronomer Giovanni Schiaparelli reported seeing a network of canals on Mars. His report was subsequently confirmed by other astronomers, and Percival Lowell described the canals as "a vision of a thread stretched across orange seas." However, not all astronomers saw them. In the drawings of Mars below, the one at left, made by Schiaparelli on the basis of telescope sightings, shows many canals clearly marked. The drawing at right, showing the same region on the planet as observed by E. M. Antoniadi, has similar markings but no hint of canals.

The canals, although sighted visually through telescopes, never appeared on photographs from the earth. Recent close-ups of Mars from passing spacecraft also show no signs of them. It is now generally believed that they were imagined by hopeful astronomers straining at the limits of visibility.

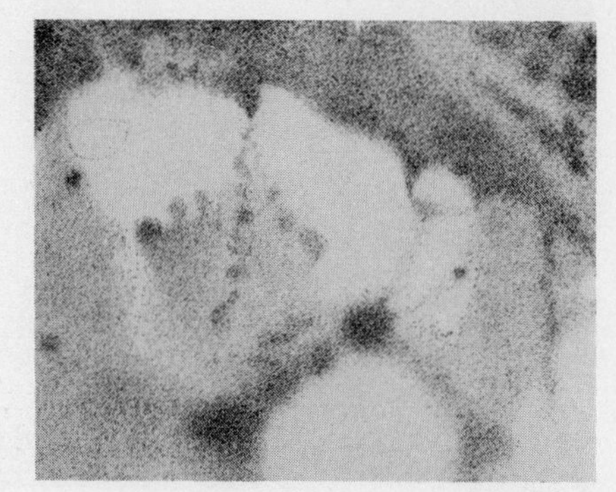

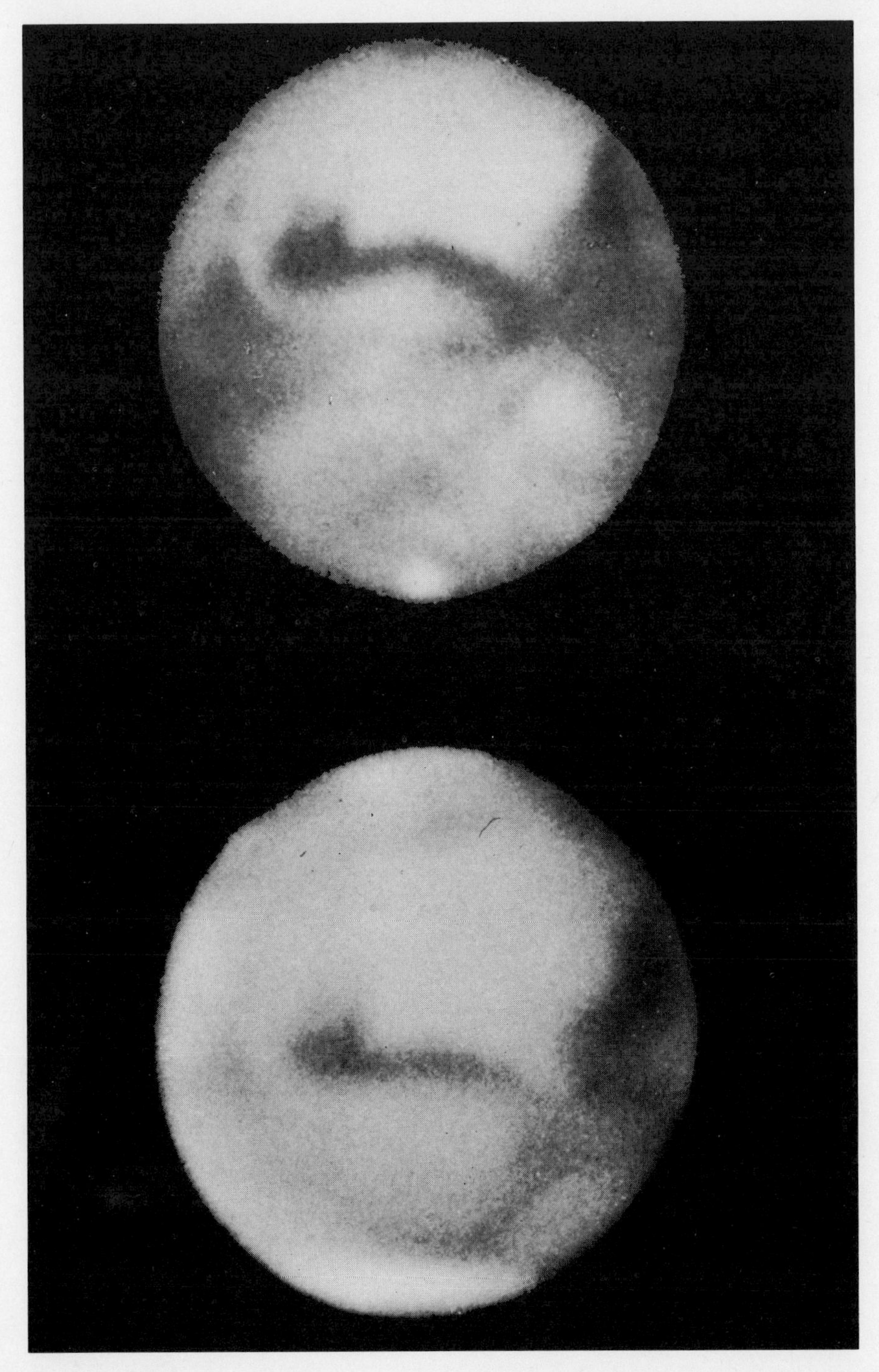

MARS PHOTOGRAPHED FROM THE MARINER SPACECRAFT: A MODERATELY WELL-PRESERVED SURFACE. Photographs of Mars, taken by Mariner spacecraft as they swept by the planet at a distance of 2000 miles, show the surface to be pitted by meteorite craters resembling craters on the moon. A mosaic of Mariner photographs (*above*) shows hundreds of craters in an area approximately 400 by 1500 miles in size. The largest crater in the mosaic is 160 miles in diameter.

Lunar craters similar to those visible here are surrounded by circular ramparts ranging up to 15,000 feet in height and contain central peaks that rise thousands of feet from the crater floor. The ramparts and central peaks are missing from many of the Mars craters, indicating stronger forces of erosion than exist on the moon. Because Mars is more eroded and timeworn than the moon, it is less promising as a source of clues to the beginning of the solar system. However, it is more promising as an abode for life. Simple, hardy plants could survive on Mars today, and a variety of living organisms may have existed there at an earlier time if the planet once had an abundance of water.

The search for life or the remains of life on Mars will be the main objective of planetary exploration in this century. The search will begin in earnest around 1975 with the landing of a small automated laboratory, containing tests for Martian microbes and for the chemical building blocks of living matter.

HELLAS

THE MYSTERY OF HELLAS. The 1969 Mariner photographs revealed an extraordinary feature in the region called Hellas, a circular area of about one million square miles in the southern hemisphere, marked on the Mars globe (*opposite*). Hellas, unlike all other areas of Mars that were photographed, is devoid of craters. The mosaic of two photographs below, covering a strip about 800 miles long on the surface of Mars, displays the transition from the normal Mars terrain to Hellas. The first photograph shows a normally cratered region called Noachis. The second shows Hellespontus, on the border between Noachis and Hellas. An enlarged view of one part of this border region, 45 by 60 miles, is shown at far right outlined in black. Only a few craters appear in this view, in the lower left corner. The region to the right beyond Hellespontus, in the interior of Hellas, is seen to be completely craterless.

Since Hellas was undoubtedly bombarded by meteorites with the same intensity as other parts of the Martian surface, the absence of craters indicates that an unusual leveling force has been at work throughout this area on Mars. The mysterious force may be connected with an unusual concentration of planetary heat and moisture in the area, conditions that favor the evolution of life.

Hellas is distinguished in another respect. It is one of the areas that display seasonal changes in color, becoming darker with every Martian spring and lighter again in the fall. Whether the darkening is connected with the unusual leveling force in this region, and whether the two circumstances together indicate the seasonal growth of plants on Hellas, must remain tantalizing speculations until further unmanned reconnaissance of the planet is carried out in the 1970s. It is possible that the mystery of Hellas will be solved only when men land on Mars.

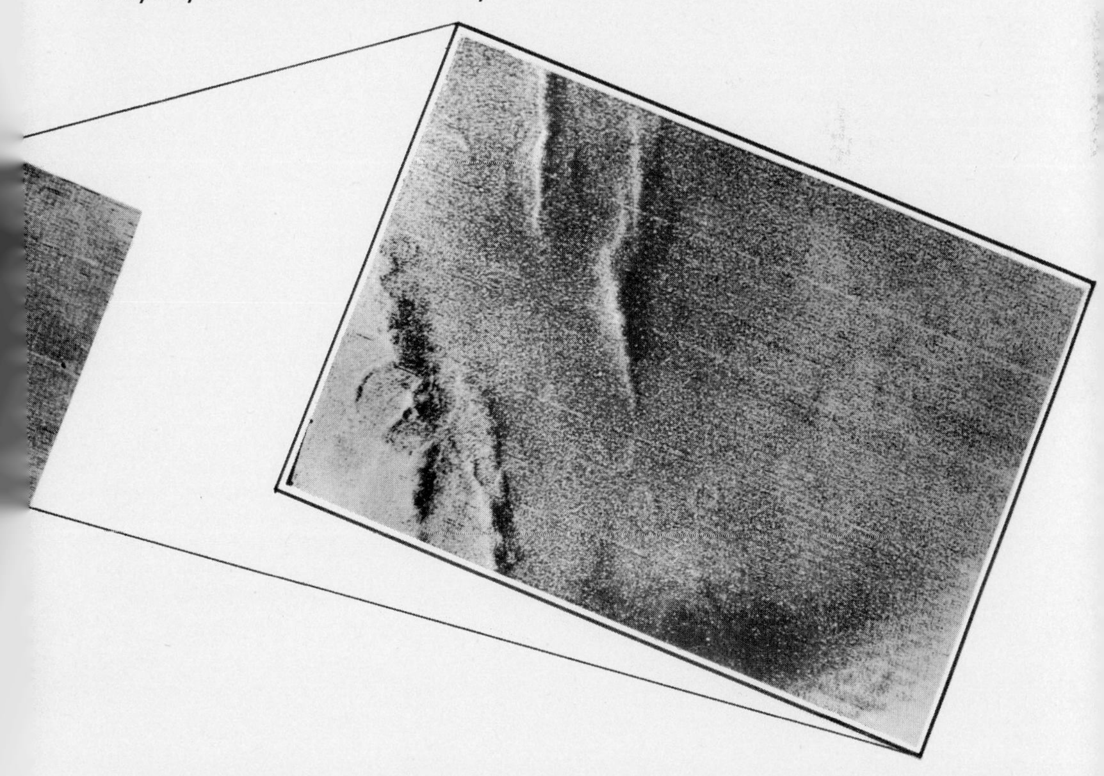

JUPITER: THE GIANT PLANET. Jupiter is the largest planet in the solar system. It completes one circuit of the sun in five years. Jupiter is 450 million miles from the sun—well outside the orbits of the earth and Mars—and receives only 4 percent as much sunlight as the earth.

Jupiter, like Venus, is covered by a thin layer of clouds (*right*). The clouds are divided into bands of varying colors and brightnesses running parallel to the equator. The cause of the colors is not known. The temperature at the level of the cloud tops is 270 degrees below zero Fahrenheit. Below the clouds the temperature must rise, and at some intermediate altitude it doubtless passes through a range suitable for life.

The dark oval-shaped region in the photograph is the mysterious Red Spot, 40,000 miles long. No satisfactory explanation of the Red Spot exists. The Spot drifts with respect to the surface of the planet, showing that it cannot be caused by a peculiarity of the underlying terrain.

One of Jupiter's moons is visible near the planet at the upper right. Its shadow appears as a dark circle on Jupiter above the Red Spot.

Jupiter's atmosphere is composed of ammonia, methane, hydrogen and helium, and probably liberal amounts of water vapor. Laboratory experiments have shown that the molecular building blocks of life—the amino acids and the nucleotides—are readily formed in this particular mixture of gases. Probably these gases were present in abundance in the atmosphere of the earth when it was a young planet and were the starting point for the evolution of life. It is possible that living organisms also have developed on Jupiter.

GRAND TOUR. In 1977 the planets will be favorably arranged for a "Grand Tour" of the outer solar system. A spacecraft sent out from the earth toward Jupiter will be whipped around by the enormous gravitational pull of the Giant Planet and will pick up enough speed from the encounter to send it farther out into the solar system, first to Saturn, and then around Saturn toward Pluto (*below*). The whole trip will take 13 years. This opportunity for a low-cost exploration of the farthest reaches of the solar system occurs only once every 179 years.

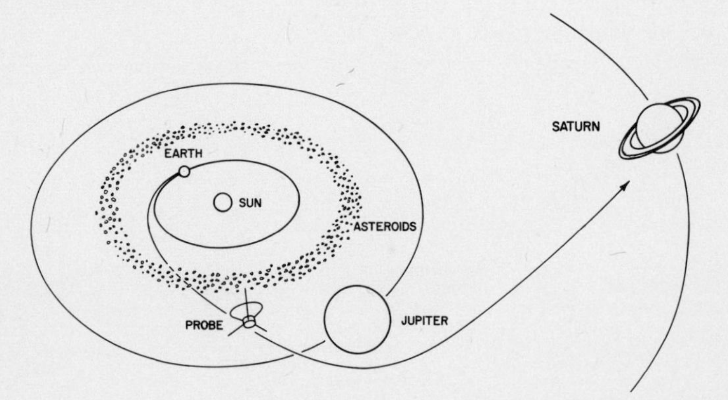

☼

9 The Age of the Earth

IN 1648 James Ussher, Archbishop of Armagh, pronounced the creation of the earth to have occurred in 4004 B.C. A span of roughly 6000 years for the age of the earth, based on the biblical genealogy, was accepted by nearly everyone until the beginning of the nineteenth century, when geologists and naturalists began to suspect that the earth must have existed for a much longer interval of time.

During the course of the nineteenth century, geologists observed that land is stripped from the continents by water erosion; streams and rivers wear away their banks, carrying a small amount of soil from the highlands to the seas every year. Although only a minute fraction of the mass of the land is removed each year in this way, apparently the process has been going on for a very long time, for there are places in which one can see that an entire mountain range has been worn down to its roots in this way. Usually it is difficult to detect these changes, but in a few special locations their effect can be seen very clearly. The Grand Canyon is the most striking example. There the Colorado River has cut through the earth's crust like a scalpel through tissue, exposing a clean record of the past. The history of that part of the earth is laid bare for everyone to see.

The cliffs of the Grand Canyon show that on no less than three successive occasions mountains have been raised to great heights by the heaving and buckling of the earth's crust, eroded away by running water, and created again in later upheavals. In other places, such as the Sheep Mountain in the Big Horn basin in Wyoming, the roots of a single mountain lie clearly exposed. The signs of erosion can be seen all around us on the face of the earth. Running water is the most effective agent, but glaciers and the sand-blasting action of dust-laden winds also take their toll. The sense of never-ending change on the earth came most strongly to James Hutton, the father of modern geology, who wrote in 1795:

> . . . from the top of the mountain to the shore of the sea . . . everything is in a state of change; the rock and solid strata slowly dissolving, breaking and decomposing, . . . the soil traveling along the surface of the earth on its way to the shore; and the shore itself wearing and wasting by the agitation of the sea.

How long has the surface of the earth been wasting away in this fashion? We can measure the amount of material which is removed each year from the American continent, for example, by the major rivers such as the Mississippi. It is estimated that 800 million tons of soil are washed away to the sea from the continental United States in a year. At this rate the level of the land is lowered by one foot in 10,000 years. A plateau two miles high is removed in 100 million years. Because this thickness of material has been worn away and replaced several times in some parts of the globe, it follows that the earth must be at least several hundred million years old. Moreover, the ceaseless cycle of erosion and uplifting has been going on as far back into history, and as deep into the crust of the earth as we are able to see. There is no clue as to where and when to stop. For all the geologist can tell, the earth may have existed forever. Hutton said: "We find no vestige of a beginning."

Several generations later, Charles Darwin undertook a study of living creatures and their relation to the fossils of ancient animals preserved in the rocks. There are now approximately one million species of animals on the face of the earth. Darwin saw that the forms of life existing on the earth today have evolved gradually out of earlier and simpler beginnings through a succession of very small changes. From the examination of a large number of fossil skeletons, it can be seen that over a period of 60 million years the modern horse, for example, has evolved from a small, five-toed animal the size of a fox terrier, as a result of a long series of minor modifications. Darwin noted that in modern animals the changes from one generation to the next are imperceptibly slight, too slight to be detectable within the lifetime of one person, or even within the memory of the human race. He concluded that a substantial change in the form of an animal must have required thousands if not millions of generations, and that a vast amount of time must have elapsed from the beginning of the fossil record to the present. Furthermore, he suggested that a long interval must have elapsed prior to the appearance of fossils, in which soft-bodied animals filled the primitive seas without leaving any trace of their existence.

Exactly how long have these changes in the forms of living creatures been going on? Darwin could not answer this question, but he, too, felt intuitively that the earth must have existed for many hundreds of millions of years.

Darwin's views regarding the antiquity of the earth clashed with the opinions of Lord Kelvin, the British mathematical physicist. Geophysical measurements had revealed that a small amount of heat steadily flows from the interior of the earth to its surface. According to Kelvin, this flow of heat resulted from the fact that the earth was a molten mass when it was newly formed. During the course of millions of years, as it lost heat from the surface, it gradually cooled and hardened. Kelvin calculated how long it would take for the earth to cool to its present temperature, assuming that heat had been flowing out of the interior of the planet throughout its past history at the same rate at which it flows out today. The calculations indicated that the earth had been cooling down for about 40 million years. Kelvin announced that this was the age of the earth. He thought that 40 million years was, if anything, a generous estimate, because when the earth was a younger and hotter planet, it probably lost heat at the surface more rapidly than it does today.

But, according to Darwin, a far greater time than 40 million years was needed for species of plant and animal life to have developed by the slow process of natural selection. If Kelvin's calculations were correct, Darwin's theory of evolution must be wrong.

Darwin was very upset by this development. In 1869 he wrote: ". . . I am greatly troubled at the short duration of the world according to Sir Thomas W. Thompson [Lord Kelvin] for I require for theoretical views a very long period before the Cambrian formation." The strain must have been grievous, for he seems to have taken a personal dislike to Kelvin, referring to him, in a letter to Alfred Wallace, as "the odious spectre." But Kelvin was confident; in 1873 he said: "We find at every turn something to show . . . the utter futility of [Darwin's] philosophy." By 1893 he had reduced his estimate of the age of the earth to 24 million years, squeezing Darwin relentlessly.

In the conflict between Darwin's intuition and Kelvin's mathematical physics, intuition triumphed, for Kelvin had omitted a major factor from his calculations in the 1870s. The missing factor did not come to light until 1904, twenty-five years after Darwin's death. In that year Rutherford discovered that radioactivity releases appreciable amounts of heat. The earth contains radioactive substances—thorium, uranium and potassium—in its interior. Kelvin had not known of the existence of these substances. According to Rutherford's measurements, they released enough heat to extend the time of cooling of the earth far beyond Kelvin's estimates.

Rutherford discussed the implications of his experiment in a lecture given in 1904, which Kelvin attended. Afterward Rutherford said:

> "I came into the room, which was half dark, and presently spotted Lord Kelvin in the audience and realized that I was in for trouble at the last part of my speech dealing with the age of the earth, where my views conflicted with his.

To my relief, Kelvin fell asleep, but as I came to the important point, I saw the old bird sit up, open an eye and cock a baleful glance at me! Then a sudden inspiration came, and I said Lord Kelvin had limited the age of the earth, provided no new source was discovered. That prophetic utterance refers to what we are now considering tonight, radium! Behold! the old boy beamed upon me."

The discovery of radioactivity released the naturalists from the strait-jacket of Kelvin's calculations and gave them as much time as they needed. It is curious that the same discovery also yielded the first indication of the actual age of the earth. The measurement was conceived, once again, by Lord Rutherford, and worked out in detail by B. B. Boltwood of Yale University. Boltwood had undertaken to repeat the work of Pierre and Marie Curie on the separation of radium from uranium. In a study of uranium-bearing rocks he isolated and measured the concentration of uranium of every other readily analyzable substance in his sample rocks. He discovered that whenever uranium was present in the rocks, lead also occurred and, furthermore, that the ratio of lead to uranium was nearly always the same.

The Curies had shown that when uranium decays, a succession of particles comes out, until finally lead appears as the end product. As the atoms of uranium decay one by one, the concentration of lead in the rock must steadily increase. The amount of lead present, Rutherford reasoned, will reveal the length of time over which the decay from uranium to lead has been occurring.

Laboratory measurements show that 50 percent of the uranium in a rock will change to lead in the course of 4.5 billion years. This is known as the half-life of uranium. Boltwood found a large amount of lead in the rocks he studied, indicating that they had existed for an appreciable fraction of the uranium half-life, perhaps for as long as a billion years. In subsequent years still older rocks were dated in this way, yielding ages ranging up to 3.3 billion years.

Meteorites, which are pieces of extraterrestrial rock that occasionally collide with the earth, have also been studied by the same technique. The results show that most meteorites are about 4.5 billion years old. Meteorites have not been worked over by erosion, and subjected to a variety of chemical and physical changes, as is the case for the rocks on the surface of the earth. For this reason, it is believed that they give a better indication of the original state of matter in the solar system than is given by terrestrial rocks. Therefore, 4.5 billion years, the age of the meteorites, is assumed to be the age of the solar system and the age of the earth.

THE GRAND CANYON: AN INCISION IN THE CRUST OF THE EARTH. The sides of the Grand Canyon show the remains of three mountain ranges which have been raised up and successively worn away to their roots by the erosive action of running water. The Colorado River *(below, running from left center to bottom right)* has cut through the crust of the earth to a depth of 6000 feet to expose the record of these changes. It reveals layers of the earth's crust which were formed as long ago as one billion years.

CHARLES DARWIN (1809-1882): DEFENDER OF AN ANCIENT EARTH. This photograph was taken in 1854, when Darwin was forty-five years old. At that time he had been thinking about fossils and their relation to living animals for nearly twenty years. He had concluded that the forms of life now on the earth have developed out of vanished species by a succession of innumerable small changes. These changes, imperceptible from one generation to the next, must have occurred over an exceedingly long period of time. Darwin was convinced that the earth was a very old planet.

LORD KELVIN (1824-1907): EXPONENT OF A YOUNG PLANET. Lord Kelvin, one of Britain's greatest physicists, disputed Darwin's views on the age of the earth. Kelvin's calculations showed that the earth had been a molten, uninhabitable body of rock no more than 40 million years ago. This interval of time was too short for life to have evolved to its present variety and complexity by a succession of small changes, as Darwin's theory of evolution proposed. Darwin died in 1882, deeply troubled by Kelvin's criticism. In 1907, twenty-five years after Darwin's death, Rutherford discovered that radioactive substances buried in the earth's interior released sufficient heat to invalidate Kelvin's calculations.

☼

10 The Early Years

THE early years of the earth's history are shrouded in mystery. Erosion by wind and running water, and the upheavals that accompany the building of great continents and mountain chains—all have combined to erase the record of the earth's past. We know less of the history of our planet than we do of the life story of the stars, for the skies contain stars of many different ages—some in the process of formation, some in their middle years, and some in the final stages of decay and dissolution. All these stars—young and old—are available for examination in our telescopes. Through them we have learned the story of the red giants and the white dwarfs. But planets of different ages are not available for inspection. The exploration of the moon and Mars will soon be accomplished, and their inspection will yield new information regarding the history of the planets; but the moon and Mars are probably the same age as the earth, and we will never learn the complete story from them; we will never learn as much as we could if we were to witness the birth of an earthlike planet, or to observe one in the early years of its life.

Perhaps we will discover a planet younger than the earth one day, if ever we are able to leave this solar system and voyage to other stars, but such voyages are not in prospect in the foreseeable future. For the present we are doomed to ignorance of the conditions which existed on the earth during its early years. We do not know the temperature at the surface of the young earth; the gases which floated in its atmosphere; and the chemicals which were dissolved in the primitive oceans. These facts are unknown, and perhaps unknowable, and yet they are of great interest, because they are bound up with the question of life's origin. The earliest traces of living organisms discovered thus far—residues of bacteria and simple plants—are found in rocks about 3 billion years old. When these organisms were alive the earth had already existed for more than one billion years. During that period of a billion years life developed here. *What were the conditions under which it arose?*

We believe that at the beginning there was only a cloud of gaseous hydrogen, mixed with small amounts of other substances. Out of this cloud grew the sun, the planets, and the creatures which walk on the surface of the earth. It was the parent cloud of us all. At its center existed a dense, hot nucleus which later formed the sun. The outer regions—cooler and less dense—gave birth to the planets.

Out of what materials were the planets formed? The bulk of the parent cloud must have been composed of the light gases, hydrogen and helium, because they are the most abundant elements in the universe. Other elements relatively abundant in the universe, although less so than hydrogen and helium, are carbon, nitrogen and oxygen, metals such as iron, magnesium and aluminum, and silicon. These substances must also have been present in relatively great abundance in the parent cloud of the planets. No doubt the remaining eighty-odd elements were also represented, but in smaller amounts.

All the familiar chemical compounds of these substances would have formed in the cloud in a relatively short period of time. Hydrogen combines readily with oxygen to form molecules of water vapor; hydrogen also combines with nitrogen to form molecules of ammonia gas, and it combines with carbon to form methane, also called marsh gas, which is used extensively today for cooking. Carbon and oxygen combine to form carbon dioxide. Considerable amounts of each of these compounds must have formed in the parent cloud. However, they were probably not present in the form of gases, because of the low temperature—about 100 degrees Fahrenheit below zero—prevailing in the region of the cloud out of which the earth was formed. At this temperature they congealed into a slushy mixture of water, ammonia and methane ice in liquid and solid form, plus solid carbon dioxide—dry ice. The other elements that were present in abundance—silicon, aluminum, magnesium and iron—combine with oxygen to form grains of rocklike materials and metallic oxides.

These, then, are the substances out of which the planets condensed: a Neapolitan sherbet of frozen water, ammonia and methane, plus various kinds of rocky substances—all immersed in a gaseous cloud of hydrogen and helium.

When the planets first condensed out of this mixture of gases and solid matter, the bulk of their mass should have consisted of hydrogen and helium. The giant planets—Jupiter, Saturn, Uranus and Neptune—are in fact composed mostly of these light gases, but for some reason the earth and its nearest planetary neighbors lack them. Why they are scarcer on the earth than on Jupiter is a mystery. Some students of the subject say that they were blasted away by the rays of the sun, which was much more brilliant than it is today. Others say the opposite: that all the light gases in the inner regions were drawn into the body of the primitive sun as it contracted, and the earth was formed out of the rocky materials left behind.

Whatever may have caused the departure of hydrogen and helium, it is clear, from the scarcity of these gases on the earth, that they had disappeared from the neighborhood of the earth's orbit to a large extent by the time our planet had started to condense. There remained only particles of rock and small amounts of ice, which circled in orbit around the sun, each a miniature planet in its own right.

Occasionally collisions occurred between neighboring particles in the course of their circling motion. Some collisions were gentle, and the particles stuck together. In this way, in the course of millions of years, small grains of rock gradually grew into larger ones. Some pieces of rock became large enough to exert a gravitational attraction on their neighbors. These were the nuclei of the modern planets. Once they had grown large enough to attract other particles by their own gravity, they quickly swept up all the materials in the space around them, and developed into full-sized planets in a short time.

The complete process of planet formation went on over a period of perhaps 50 million years, proceeding with extreme slowness at first, and then with rapidly increasing momentum in the final stages. At the end, all the matter of the solar system was gathered into the existing planets, and only a few atoms of gas remained in the space between. This is the situation in the solar system as it exists today.

As the earth grew to its final size in the last stages of this process, the forces of its gravity drew down on the planet, with increasing force, all the pieces of rock that still circled the sun in neighboring orbits. These last remnants of the parent cloud plunged into the earth at speeds up to 25,000 miles per hour, liberating great amounts of energy as they hit the surface, and raising the temperature of the earth's outer layers.

It is possible that the entire earth melted as a result of bombardment suffered during the final stages of its formation. Or it may have melted later on as a result of the nuclear energy released in its interior by the decay of uranium and other radioactive substances. There is a dispute among students of the earth's history on this point, but of one fact we are certain: Large parts of the earth melted, or came very close to melting, at some point in its history.

We are sure of this because at the present time there is a very large amount of molten iron at the center of the earth. Iron is an abundant element, and we are not surprised to find a considerable amount of it in the interior of the earth; however, when the earth first accumulated out of small pieces of rock, this iron must have been sprinkled randomly throughout its interior, like raisins in a fruitcake. Only if large parts of the earth had melted would it have been possible for this iron to run to the center and form a molten core.

The question is, when did the earth melt? The "cold young planet" school of earth history argues that the earth was initially a cold body, composed of solid rock, which melted subsequently, perhaps a billion years later, by the heat released from radioactive substances distributed throughout its interior. The "hot young planet" school asserts that the earth was molten at the beginning of its existence, and did not form a solid crust until considerably later.

One way or the other, it is certain that large parts of the earth were melted at some point during its first billion years or so. Gradually, light rocks accumulated at the surface to form the continents. The areas between the continents were natural basins in which water, reaching the surface from the interior of the planet through volcanoes and fissures in the crust, collected to form the oceans. Slowly the earth acquired its present appearance.

11 The Dawn of Life

THE earth began its existence 4.5 billion years ago, circling around the newborn sun. Formed out of inert atoms of gas and grains of dust, our planet was surely a sterile body of rock at the beginning. The waters of the primitive oceans were devoid of life; their waves lapped at barren shores, uncarpeted by vegetation. Yet today plants grow everywhere; the continents crawl with a million varieties of animal life; 20,000 kinds of fishes inhabit the seas. How and when did this rich variety of living forms appear on our planet?

Advances in science in recent decades have uncovered facts about the nature of living organisms which lead for the first time to a scientific explanation of the origin of life. It now appears likely that the first living creatures on the earth evolved spontaneously out of the inert chemicals that filled the atmosphere and oceans of the planet in the early years of its existence. Three discoveries lead to this conclusion.

First, biologists have shown that all living organisms on the face of the earth depend on two kinds of molecules—amino acids and nucleotides, which are the basic building blocks of life—just as the physicists have shown that all matter in the universe is constructed out of three building blocks—the neutron, the proton and the electron.

Second, chemists have manufactured these molecular building blocks of life in the laboratory out of simple chemicals, under conditions resembling those that existed on the earth when it was a young planet.

Third, an object has been discovered which links the nuclei, atoms and molecules of the physical universe to the complex organisms of the living world. This object, called the virus, lies on the borderline between inanimate matter and life. Its existence gives credibility to the notion that life evolved out of nonliving chemicals.

The basic building blocks of life are more complicated than the building blocks of the physical world. Twenty different kinds of amino acids play a critical role in living creatures, and five different kinds of nucleotides.*

* It is more correct to say that there are five *nucleotide bases.*

Furthermore, each amino acid or nucleotide is, itself, a rather complex molecule made up of approximately thirty atoms of hydrogen, nitrogen, oxygen and carbon, bound together by electrical forces of attraction. Examples of the structure of a typical amino acid and a typical nucleotide are shown in the illustration on page 129.

The amino acid and the nucleotide have very different functions in the chemistry of life. Within the cell the amino acids are linked together into very large molecules called proteins. One class of proteins, called structural proteins, makes up the structural elements of the living organism—the walls of the cell, hair, muscles and bone. The structural proteins are like the steel framework and walls of a building. The other type of protein is called the enzyme. Many kinds of enzymes exist; each kind controls one of the many chemical reactions that are necessary to sustain the life of the organism.

All proteins in all forms of life, plant and animal, are constructed out of the same basic set of twenty amino acids. One protein differs from another only in the way in which its constituent amino acids are linked together. However, these differences are all-important. The distinction between a man and a mouse, in both appearance and personality, depends entirely on the differences between the proteins contained in the cells of their bodies.

Proteins are assembled within living organisms by the second set of building blocks—the nucleotides. Nucleotides are joined together within the cell to form very long chains, called nucleic acids. The most important type of nucleic acid is called deoxyribonucleic acid, or DNA for short.* DNA is the largest molecule known, containing, in advanced organisms such as man, as many as 10 billion separate atoms. The size of the DNA molecule is understandable when we consider the complexity and importance of its functions in the living cell. The DNA molecule is the most important molecule in every living organism, even more important than the protein, because it determines *which* proteins will be assembled; the DNA molecule has the master plan for the organism.

How does DNA control the assembly of proteins in the cell? The general features of the process began to emerge during the decade of the 1950s, although many of the details are still not clearly understood. It appears that the separate amino acids and nucleotides float freely in the fluid of the cell. The DNA molecules which direct the assembly of proteins are located in the

* Only four of the five important nucleotides enter into the structure of DNA. The fifth nucleotide belongs to another type of nucleic acid.

center of the cell. In the first step, free nucleotides are attracted to a segment of one of the DNA molecules at the center. They line up alongside the DNA segment to form a replica of it. In the second step, the replica detaches itself from the master DNA chain, and drifts off into the cell; it is a messenger which carries instructions from the DNA into the body of the cell for the assembly of one particular kind of protein. In the third step, another molecule enters the picture. This molecule serves as a connecting link, bringing the amino acids in the fluid of the cell to the appropriate places alongside the messenger. There are twenty kinds of connecting links, one for each kind of amino acid. Each of the connecting links attracts one and only one of the 20 amino acids. When, in the course of chance collisions, the right kind of amino acid comes into contact with the end of the connecting link designed for that particular amino acid, it is held fast there. At the other end of the connecting link is another set of molecules, making up a surface of nooks and crannies so constructed that it can fit only into the appropriate place along the length of the messenger. When the connecting link takes its place along the messenger, it adds the amino acid to the chain of amino acids that has already been built up. When a chain of amino acids has been assembled along the full length of the messenger, the assembly of amino acids into a protein is complete. The assembled chain then detaches itself from the messenger and drifts off into the fluid of the cell.

By this rather complicated process, the essential proteins are built up within an animal in accordance with the order of the nucleotides in its DNA molecules. The segments of the DNA molecule are "read" like the words of a book. Each DNA segment, controlling the assembly of one protein, is a word; each nucleotide within a segment is a letter; the order of the letters provides the meaning of the word—that is, the protein to be assembled. The full set of DNA molecules contained within a cell is the library of genetic information for the organism. The DNA molecules in the cells of a human being direct the assembly of the amino acids in the human body into human proteins; the DNA molecules in the cells of a mouse direct the assembly of its amino acids into mouse proteins.

How is the plan for the assembly of the right kind of proteins passed from one generation to the next? How do progeny inherit their characteristics from their parents? The answer lies in a most extraordinary property of the DNA molecule—the ability to make a copy of itself. The mechanism by which DNA copies itself was discovered in 1953 by an Anglo-American team, James D. Watson of Harvard University and Frances Crick of Cambridge University.

Francis Crick

James D. Watson

This discovery is one of the most important single scientific events of the twentieth century to date. I have described the DNA molecule as a chain of nucleotides; but Watson and Crick found it is not a *single* chain of nucleotides; it consists, rather, of *two* chains, joined together at regular intervals by molecules which run between them like rungs of a ladder. At the middle of each rung of the ladder is a weak spot, which is easily broken. During the early history of a cell the two strands remain connected, but when the cell has attained its full growth, and the division into two daughter cells is about to commence, the weak connections running down the middle of the ladder break, and the double strand separates into two single strands. Each of the single strands then collects unto itself new nucleotides out of the pool of nucleotides floating in the cell, and assembles them into a new double-stranded molecule. There now exist two identical DNA molecules, where formerly there was one. The two DNA molecules separate, and move to opposite corners of the cell; the cell then divides into two daughter cells, each containing one complete set of DNA molecules. Thus, each of the daughter cells contains a copy of the volume of genetic information which had been in the parent cell. This is the way in which the shape and the character of a plant or an animal are transmitted from generation to generation.

In summary, the DNA molecule controls the assembly of proteins, and the proteins determine the nature of the organism. Each living organism has its own special set of DNA molecules; no two organisms have the same set unless they are identical twins. However, *the basic nucleotides and amino acids are the same in every living creature on the face of the earth, whether bacterium, mollusc or man.*

With this fundamental property of living creatures in mind, one can appreciate the importance of a critical experiment performed in 1952 by Stanley Miller. Miller, now a professor of biochemistry at the University of California, was then a graduate student working on his Ph.D. thesis under Harold Urey. At the suggestion of Urey, Miller mixed together the gases—ammonia, methane, water vapor and hydrogen—which were abundant in the parent cloud of the earth, and which were probably abundant in the earth's primitive atmosphere. He circulated the mixture through an electric discharge. At the end of a week, Miller found that the water contained several types of amino acids. Subsequently, in 1962, nucleotides were created in the laboratory under similar conditions. In similar experiments amino acids and nucleotides have been manufactured out of a variety of gas mixtures, using various sources of energy—bombardment by alpha particles, irradiation with ultraviolet light, and simple heating of the ingredients. The results of all these experiments, taken together, demonstrate that the molecular building blocks of life could have been created in any one of many different ways during the early history of the earth.

Amino acids and nucleotides might have been formed on the earth in this way 4.5 billion years ago, by the discharge of lightning in primitive thunderstorms, or by the action of solar ultraviolet rays from the sun. We can guess what happened thereafter. Gradually the critical molecules drained out of the atmosphere into the oceans, building up a nutrient broth of continuously increasing strength. Over a long period of time the concentration of amino acids and nucleotides increased, until eventually a chance combination of building blocks produced still more complex molecules—primitive proteins and nucleic acids. With the further passage of time, cells developed; many-celled organisms appeared; and living organisms were started on the long road to the complexity of the creatures which exist today.

This is a nice story that has emerged from the union of astronomy, biology and chemistry. Yet, it is hard to believe that existing forms of life, in all their variety and sophistication, can be traced back to simple chemicals. Is there any direct evidence for the development of life out of nonliving molecules?

The answer is: Yes, there is an entity, very common in the world today, which possesses, at the same time, the attributes of a nonliving molecule and the attributes of a living organism. This entity is the virus—the smallest and simplest object which can be said to be alive.

The existence of viruses first came to light at the end of the nineteenth century, in the course of a series of experiments designed to reveal the cause of a disease affecting tobacco plants. It was found that the juice pressed from the leaves of infected plants could transmit the infection to other plants. Apparently, the infection was transmitted by bacteria contained in the fluid. But when pressed through a fine filter, which screened out all visible bacteria, the fluid still retained its power of infection. In 1898 a Dutch botanist, Beijerinck, suggested that the disease was not caused by a germ, but by a poisonous chemical. Beijerinck called the chemical a "virus," which is the Latin word for poison.

Further research revealed that viruses are the cause of many diseases, including smallpox, influenza, infantile paralysis and the common cold. Medical and biological interest in viruses intensified during the first decades of the twentieth century. Gradually, the suspicion developed that the virus was no ordinary chemical. A variety of experiments suggested that the virus, although too small to be seen under the microscope, possessed the basic attribute of living organisms—the ability to reproduce itself.

Still, the evidence for the living virus was indirect; no one had yet seen one in the act of reproduction. But in the years after the Second World War a new instrument was perfected, which provided the biologists with a powerful tool for the study of small organisms. This instrument was the electron microscope. Ordinary microscopes, in which the object under study is illuminated by rays of light, are limited to a magnifying power of approximately 2000. The smallest bacteria, which are a hundred-thousandth of an inch in size, can just barely be seen in these microscopes. But the electron microscope, which directs a beam of electrons at the object instead of a beam of light, can produce magnifications as high as several hundred thousand diameters. It is possible to photograph a single protein molecule with these instruments; if a further improvement in magnification by a factor of 50 can be achieved, the electron microscope will be able to photograph individual amino acids and nucleotides.

Under the electron microscope the virus finally became visible, and all the important details of its structure were revealed. It was found that viruses come in many shapes—round, cylindrical, polyhedral and with tails. They also come in many sizes. The largest is as large as a small bacterium; the smallest, which is a millionth of an inch in diameter, is smaller than many nonliving molecules. Viruses bridge the gap in size between the inanimate and the animate worlds.

Yet these tiny particles are indisputably alive. Chemical studies show that they contain DNA—the molecular blueprint of life and the means by which every living creature reproduces itself. They also contain a substantial amount of protein, in the form of a protective coat wrapped around the precious, delicate strands of DNA. But they contain very little else. In particular, they have none of the sugar and fat molecules that provide energy for the chemical reactions in other living creatures. They also lack free nucleotides and amino acids, out of which all other organisms make proteins and assemble copies of themselves.

How, then, without a source of energy, and without the materials essential for growth and reproduction, do viruses live?

The answer is clearly revealed by the electron microscope. A virus, by itself, is not alive. If a solution of virus particles is carefully dried out the viruses will stick together in a symmetric pattern to form a crystal as geometric—and as lifeless—as a crystal of salt or a diamond; left undisturbed, the crystal remains inert for years. But, dissolved again in water, and placed in contact with living cells, the molecules of the crystal spring to life; they fasten to the walls of the cell, dissolve a small opening in the cell wall, and, through the opening, inject their DNA into the cell. Once inside the cell, the virus DNA seizes control, displacing the original DNA of the cell and establishing itself as the master of all further chemical activity. The full molecular resources of the invaded cell—the energy-giving fats and sugars, the amino acids and the nucleotides—are commandeered and employed in the assembly, not of the proteins needed by the invaded cell, but of the proteins needed by the virus. At the same time, the virus gathers the free nucleotides floating in the cell fluid and assembles them, not into copies of the DNA of the invaded cell, but into copies of itself. The virus even secretes an enzyme which breaks down the existing DNA within the cell into its component nucleotides, in order to have more of these precious units available for making replicas of itself.

When several hundred protein coats and virus DNAs have been assembled, the cell is milked dry. The coats wrap around the virus DNA molecules to form complete viruses, while the original virus, in a final step, secretes an additional enzyme which dissolves the cell walls. An army of virus particles marches forth, each seeking new cells to invade, leaving behind the empty, broken husk of what had been, an hour before, a healthy, living cell. The operation is simple, ruthless and effective. It is executed by an organism which is, in the smallest viruses, only 200 atoms wide. The virus is truly the link between life and nonlife—the bridge between living and nonliving matter.

MOLECULAR BUILDING BLOCKS OF LIFE. Twenty distinct kinds of amino acids and five kinds of nucleotides are basic to all life on earth; they are the molecular building blocks of living matter.

Within the cell the amino acids are joined together to form larger molecules called proteins, with several hundred amino acids in each protein. The upper photograph on the opposite page shows a model of three amino acids linked into a short segment of a typical protein. Each ball in the model represents one atom.

Proteins divide into two classes: One type, the structural proteins, make up the structural elements of living organisms, such as cell walls, hair and muscles. The other type are called enzymes; they quicken the pace of life by speeding up the assembly of small molecules into larger ones within the cell.

Nucleotides are also joined together within the cell, forming long strands of a giant molecule called deoxyribonucleic acid, or DNA for short. (Four nucleotides form DNA; the fifth occurs in a related molecule.) The lower photograph on the opposite page shows a model of two pairs of nucleotides joined together to form a short segment of the DNA molecule. DNA is the molecular storehouse of genetic information: the order in which the different nucleotides are arranged along the DNA molecule determines which proteins will be assembled out of the basic amino acids; thus DNA controls the production of proteins; the proteins in turn control the nature of the organism. DNA is the most important molecule in the cell.

DNA and proteins differ from one organism to another, but the basic amino acids and nucleotides are identical in all living forms on the planet earth.

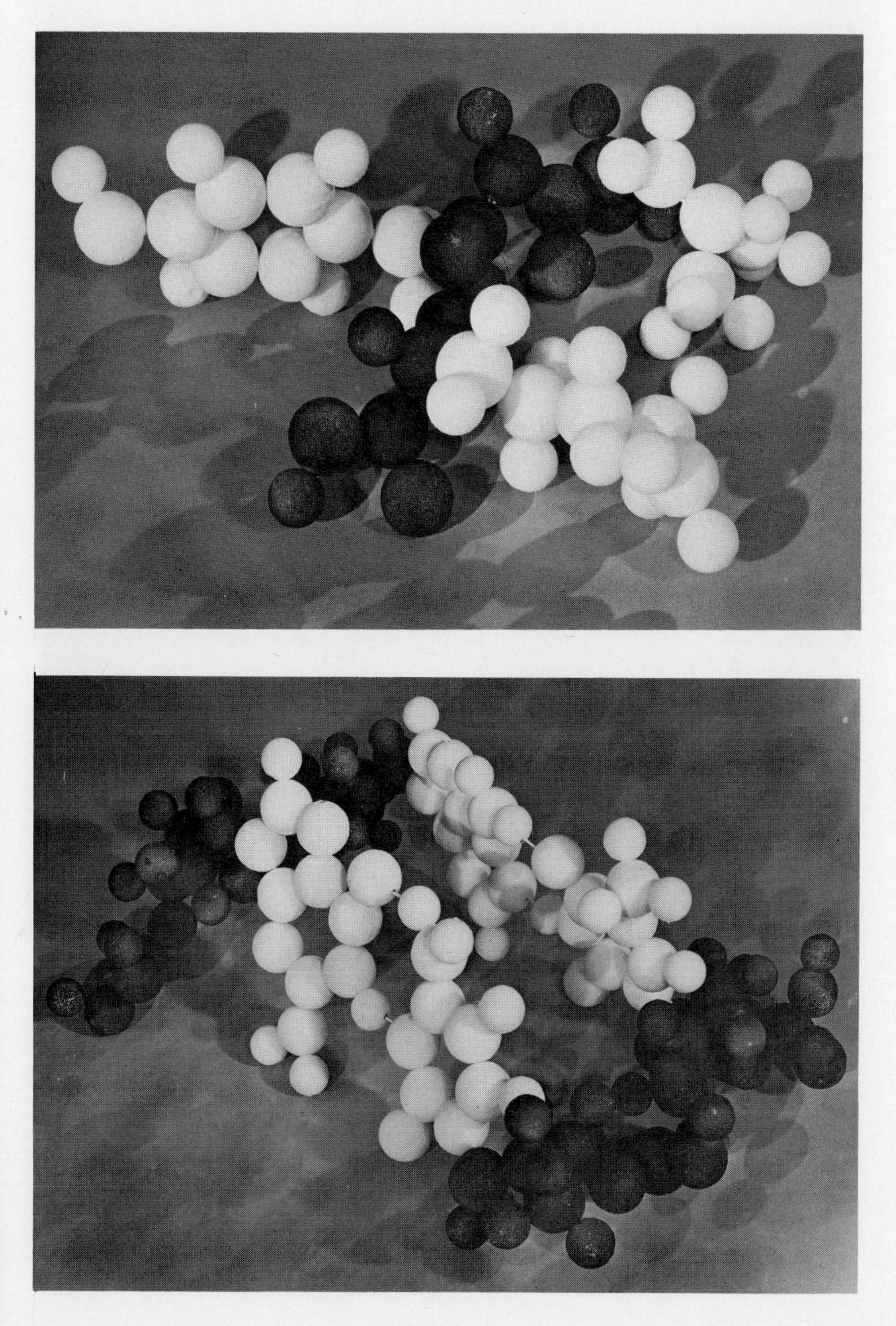

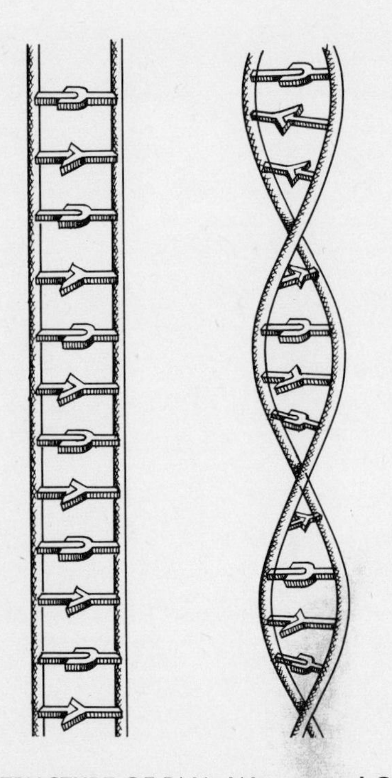

THE STRUCTURE OF DNA. Watson and Crick discovered the structure of the DNA molecule in 1952. DNA resembles a ladder *(above, left)*; each rung of the ladder is a linked pair of nucleotides, represented by the symbols

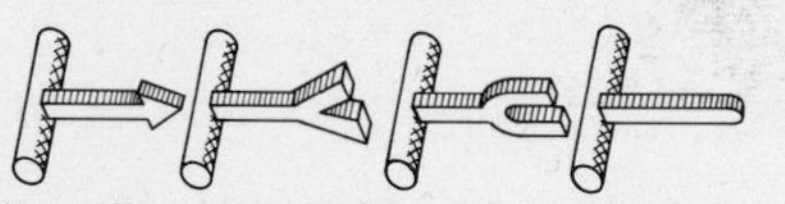

Actually, the ladder is twisted into a double spiral *(top, right)*.

The model at right shows the manner in which the double spiral of DNA is constructed out of individual atoms. Each plastic ball in the model represents one atom. This short segment of DNA contains approximately 1000 atoms; in the higher animals, the complete molecule may contain 10 billion atoms. In the photograph the author *(far right)* questions Dr. Gordon M. Tomkins, Chief of the Division of Molecular Biology in the National Institutes of Health, regarding the structure of DNA.

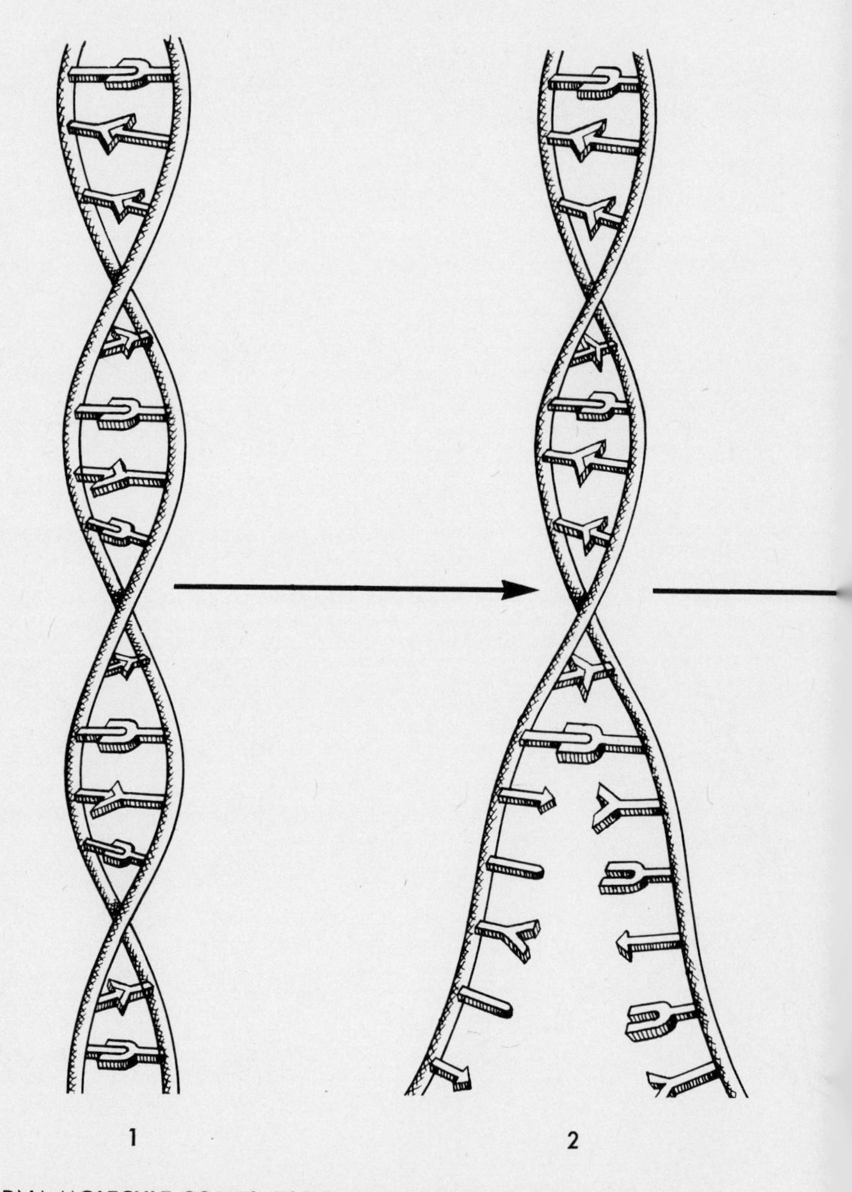

THE DNA MOLECULE COPIES ITSELF. When Watson and Crick discovered the "twisted ladder" construction of DNA, it became clear how the traits of an individual are passed from generation to generation.

The record of individual traits is stored in the DNA molecules (1), which reside in the cells of each organism. In every organism more advanced than the lowly virus, growth and reproduction involve the division of these cells. When a cell is about to divide, its DNA molecules unzip, the "ladder" untwists, the rungs break at their midpoints, and the DNA begins to separate into two parallel strands (2).

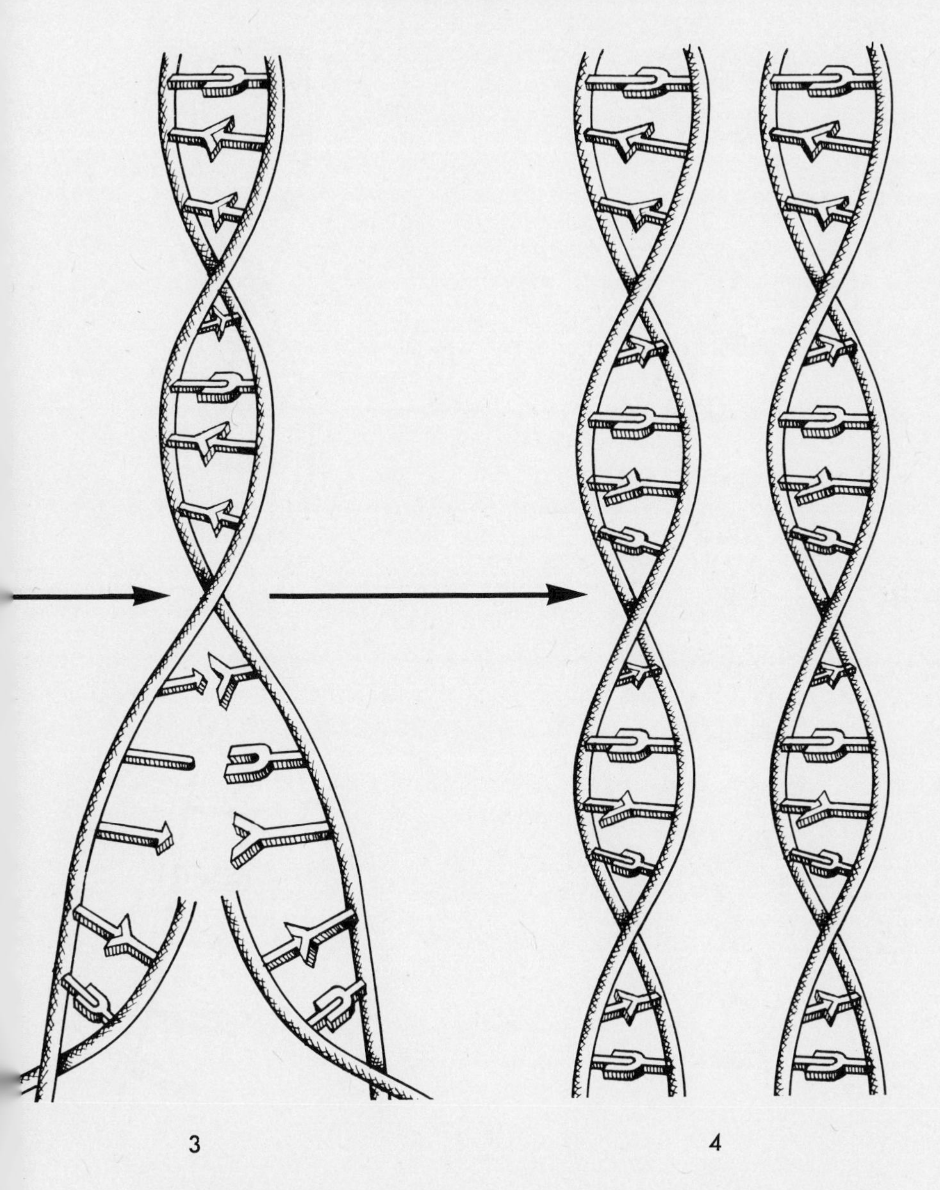

Throughout this process, unattached nucleotides float nearby in the fluid of the cell. In the next step, each strand of the unzipped DNA collects new nucleotides from the molecules surrounding it in the cell to form a new, complete "ladder" (3).

The result is two "ladders," replicas of the original DNA (4). Each replica moves to one of the two daughter cells formed in the process of cell division. In this way the master plan of the organism is passed from cell to cell and from generation to generation.

A CRITICAL EXPERIMENT: THE BUILDING BLOCKS OF LIFE CREATED IN THE LABORATORY. Biochemists have manufactured the molecular building blocks of life in the laboratory out of simple ingredients. The first experiment of this kind was performed by Stanley Miller in 1952. He mixed together gases—ammonia, methane, water vapor and hydrogen—which were probably present in the earth's primitive atmosphere, and circulated them through a glass bowl containing an electric discharge. At the end of a week, Dr. Miller found that the water contained several varieties of amino acids. Subsequent experiments have created other molecular building blocks of life in the laboratory out of a variety of chemicals and under many different circumstances. Life may have developed on the earth out of such molecules, 3 or 4 billion years ago.

The drawing of the Miller apparatus *(below)* shows the round glass bowl in which amino acids were created when a spark passed between two electrodes. At right, Dr. Miller stands next to the original apparatus with which he performed his critical experiment.

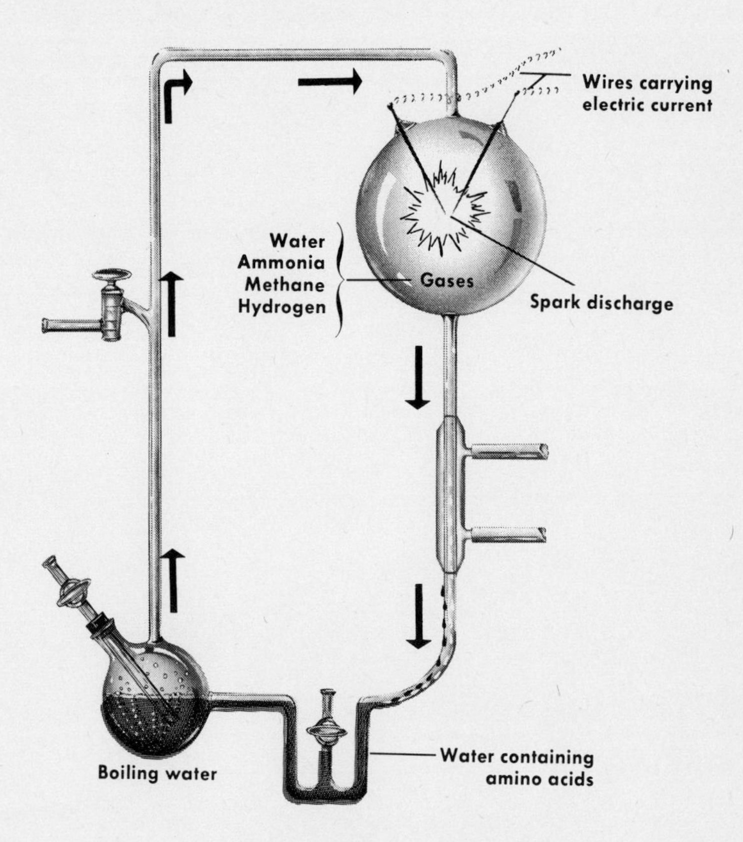

THE VIRUS: LINK BETWEEN LIFE AND NON-LIFE. The existence of the virus adds credibility to the notion that life evolved out of nonliving chemicals: the virus lies on the threshold between living matter and inanimate molecules; it is the simplest and smallest living particle, some viruses being only one-millionth of an inch in diameter.

At the left are particles of the influenza virus, magnified 100,000 times in an electron microscope.

Carefully dried out, a solution of virus particles forms completely inert crystals *(below)*. Dissolved in water and given access to living cells, the viruses composing the "crystals" come to life and attack their hosts.

On the opposite page, viruses attack a sausage-shaped bacterium. Note the minuteness of the viruses relative to the bacterium. In the lower left corner are the remains of a bacterium which was attacked one hour earlier. Viruses entered this bacterium and consumed its chemicals in making replicas of themselves, leaving the dry husk of a previously healthy cell.

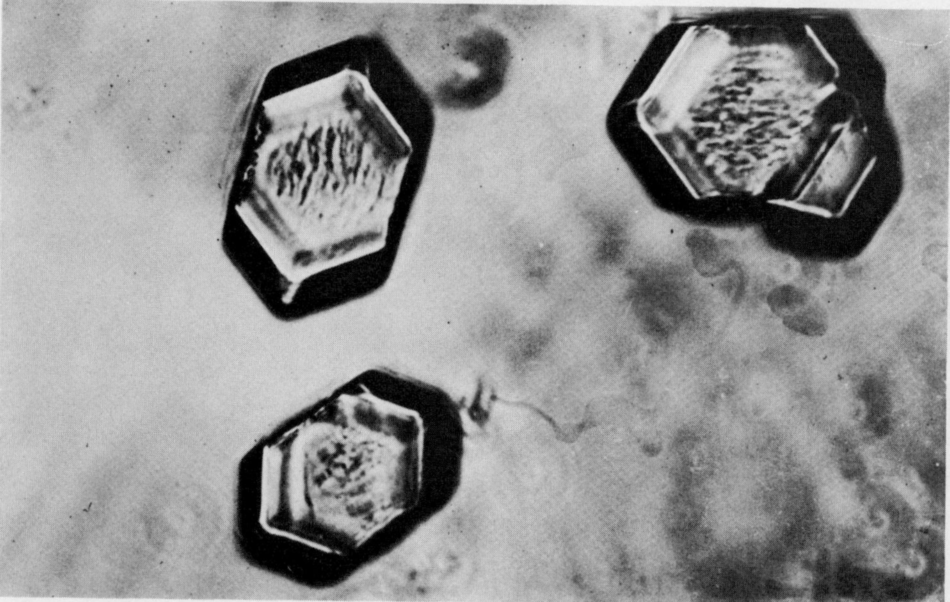

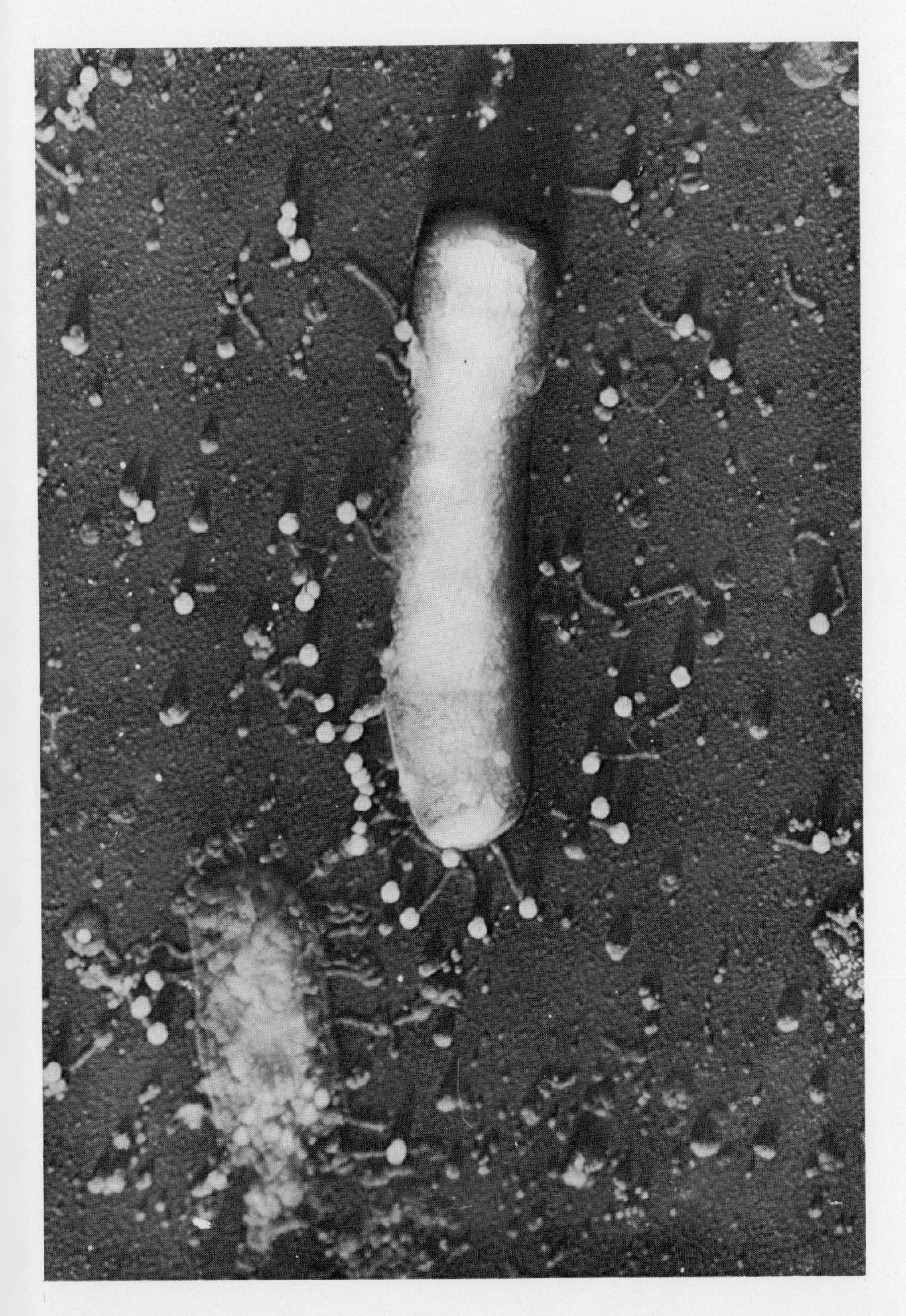

☼

12 A Message

THE imagination of the scientist has seized on these bits and pieces of research accumulated in many different fields of science, and fashioned out of them a picture of the origin of life on the earth. No living form existed on our planet in its infancy; the atmosphere was filled with a noxious mixture of ammonia, methane, water and hydrogen; peals of thunder rumbled across the sky; flashes of lightning occasionally illuminated the surface, but no eye perceived them; minute amounts of amino acids and nucleotides were formed in each flash, and gradually these critical molecules accumulated in the earth's oceans; collisions occurred between them now and then, linking small molecules into larger ones. During the course of a billion years the concentration of complex molecules increased; eventually a complete DNA chain appeared. Thus the threshold was crossed from inorganic matter to the living organism.

According to this story, life can appear spontaneously in any favorable planetary environment, and evolve into complex beings, *provided vast amounts of time are available.*

How much time is needed? Studies of the fossil record suggest that life appeared sometime during the first billion years of the earth's history. Apparently one or two billion years or so is the length of time required.

Our knowledge of the life cycle of a star indicates that the necessary period of several billion years will be available for the chemical evolution of life on any planets that circle around a star similar to our sun. Stars larger than the sun burn out too quickly to provide the needed time. Stars smaller than the sun are suitable, provided they have planets close enough to them to raise their temperatures to a comfortable range. All stars the size of the sun, surrounded by one or more planets that are approximately the same distance from them as the earth is from its star, should certainly provide very favorable circumstances for the development of living organisms.

In the multitude of planets accompanying these stars there must be some which closely resemble the earth. Let such planets be relatively few in number; let them be as rare as one in a million; no matter, the number of earthlike planets will still be 100,000 *in our galaxy alone.*

Can we maintain our belief in the uniqueness of life on the planet earth in the face of these numbers? We can on the astronomical evidence alone, because all earthlike planets except ours could be dead bodies of rock; but the biological discoveries described in the last chapter suggest that this is not the case. First, all life on the earth depends on a few basic molecules, and these molecules have been created out of simple atoms in the laboratory; second, the atoms which compose the basic molecules of life are the same as the atoms which exist on every other star and planet in the universe; third, there is reason to believe that the same laws of physics and chemistry apply in every corner of the cosmos. Therefore, the chain of physical and chemical reactions which led to the appearance of life on the earth may also have occurred on other planets.

This chain of reasoning leads us to the conclusion that forms of life resembling ours may have developed on other earthlike planets in our galaxy, and on planets in other galaxies as well. If life has developed, what is the likelihood that it is intelligent life? Is there a chance that extraterrestrial beings have acquired a level of intelligence equal to, or greater than, ours?

Considering this question, we must reflect that our galaxy is believed to be about 10 billion years old. The earth is roughly 5 billion years old and, therefore, was formed when the Galaxy had already existed for 5 billion years. Thus there must be many stars in the Galaxy that are billions of years older than the sun. Around some of these older stars circle earthlike planets on which life may have evolved. If so, this life has existed for billions of years longer than life on the earth. When we reflect on the scientific advances of the last 20 years, we realize that the advances which will occur in another billion years are beyond our imagination. Consider the history of man: we have existed as a human species for barely two million years; modern science is only 300 years old; our ability to communicate over long distances by radio goes back only 60 years; and it is a mere decade since we acquired the means of traveling in space. The period in which our scientific knowledge has developed is an exceedingly narrow slice of time, sandwiched between the billions of years of evolution that preceded the emergence of man and the billions of years that lie before us in the lifetime of the solar system.

It is exceedingly unlikely that any society on another planet came into existence at the same moment of time, and developed at the same rate, so as to have arrived at precisely the same level of technology which we possess on the earth today. A difference of 100 years, which is the blink of an eye in the lifetime of a star or planet, has produced enormous changes in the scientific knowledge of our society. Some of these extraterrestrial societies must be primitive in comparison with us; others, with an earlier start, must have surpassed our achievements a long time ago.

It is to this latter group—the more advanced societies—that we should direct our attention, for we must expect that they will have mastered the techniques of radio communication, and harnessed the power required for transmitting signals over great distances, with a greater skill than we can hope to achieve in this century. We must expect that these older, more advanced societies will reach us before we discover them.

Where are all these people? Why haven't they appeared? Regrettably, direct physical contact with societies on planets circling other stars seems an unlikely prospect in the foreseeable future, for the stars are very thinly scattered in the sky, the average distance between them being 30 trillion miles. It would take a spacecraft, moving at a rocket speed of five miles per second, about 100,000 years to cover this distance. At the present time we have no way of accelerating spacecraft to the speed required for interstellar travel.

However, interstellar *communication* is within the realm of possibility. The threshold of radio communication, which we crossed only sixty years ago, surely has been crossed on other planets thousands, if not millions, of years ago. We must expect that others who have capabilities for radio communication far in advance of ours are already listening and will hear us first. Only in the last few decades have we begun to emit enough radio and television noise to attract their attention. Sometime, perhaps soon, we may expect a message.

☼

13 Millions of Generations

MANY planets revolve about other stars; there may be millions in our galaxy, and perhaps an infinite number in the universe. Doubtless, most are dead bodies of rock, washed by sterile seas. But on some, situated at favorable distances from their suns, the environment is suitable for the formation of nucleotides and amino acids—the building blocks of life. On these planets the chaotic succession of inanimate processes gives way to a pattern of chemical evolution, complex and self-reproducing.

Life appeared on the earth as the product of this sequence of events, at some point in the first billion years of its existence. The earliest organisms were very simple, scarcely more than giant molecules immersed in the primeval waters of the planet. During the billions of years that followed, those organic molecules developed into the rich variety of plants and animals that now live on the earth. What guided the course of evolution on this planet from the first primitive organisms to the complicated creatures of today? If life has arisen elsewhere, what guides the course of evolution on other planets? Is there a law in nature which controls the forms of life?

The fossil record contains clues to the resolution of this question. Thousands of skeletons and fossil remains mark the path by which life climbed upward from its crude beginnings. The initial steps along that path are not known; those first forms must have been fragile, for no trace of them remains. The earliest signs of life to appear in the record, already far advanced beyond the "living" molecule, are the deposits of simple one-celled plants called algae, and the shells of rod-shaped organisms resembling bacteria. These are found in rocks formed 3 billion years ago, when the earth was already more than a billion years old.

Very little happened for several billion years thereafter; at least, very little that has been preserved in the record of the rocks. But suddenly, 600 million years ago, the pace of evolution quickened. In rocks of that age the first hard-bodied animals—corals, starfish, snails and trilobites—appear in great numbers. During the next 200 million years life exploded into a profusion of different forms. By 400 million B.C. all the major branches of the animal kingdom had developed.

At that time the highest forms of animal life were still confined to the waters of the planet, and the land was relatively barren. But 350 million years ago one class of aquatic animals—the fishes—developed air-breathing forms; some air-breathing fishes evolved into amphibians—the first vertebrate animals* to venture onto the land—and out of the amphibians, 50 million years later, came the reptiles. The reptiles were the first vertebrates to be completely emancipated from the water. Branches of the reptiles gave rise to the snake, lizard, turtle and bird; other branches produced the dinosaurs and their descendants, the crocodile and alligator; still other branches led to the mammals.

The dinosaurs ruled the earth for 100 million years, and during the long reign of these highly successful reptiles the mammals were kept in check and made little evolutionary progress. But suddenly, 70 million years ago, the dinosaurs disappeared. With their disappearance the mammalian stock flowered in a variety of forms, until, by 10 million B.C., the ancestors of most of the animals we see on the earth today, from aardvarks to zebras, had evolved. Two or three million years ago, late in this story, an animal recognizably similar to man appeared on the scene.

The record of these changes contains many gaps, but the segments which are present convey a clear message: Man has evolved slowly, during some billions of years, out of lower and simpler organisms. And, although the first part of the record is missing, it is probable that these lower organisms, in turn, came from nonliving molecules formed in the waters of the primitive earth.

In this book I have shown how the basic forces of nature—gravity, electromagnetism and the nuclear force—acting on the basic building blocks of matter, have led, first, to the synthesis of the elements within the interiors of stars; then, to the formation of the sun and planets out of those elements; and, finally, on the surface of one of these planets, to the formation of organic molecules lying on the threshold of life. I have shown how that threshold may have been crossed in the early years of the earth's existence. Throughout this long history my viewpoint has been that of the physicist, seeking to understand the essence of the world around him in terms of a few simple principles. One might call them the laws of physics. These laws are the distillation of all the observations regarding the physical world which have

* A vertebrate is an animal possessing a backbone and an internal skeleton, as distinct from invertebrates such as the insect, for example, whose skeleton is external and surrounds its body.

been acquired in thousands of years of human experience.

Now we have come to the explanation of the subsequent course of events in the history of life, leading from the first simple organisms to man. Here, for the first time, the principles of physics are no longer helpful. The stars and planets have yielded the secrets of their history to the physicist; the molecular foundations of living organisms are beginning to be understood; but the *complete* organism—even of the simplest and most primitive kind—is incalculably more complicated than any star, or planet, or giant molecule. New insights are needed for the understanding of its structure and evolution. A new law must be found.

The new law was discovered by Charles Darwin more than a century ago. Darwin showed that evolution is the result of a mechanism or "force" in nature, which works on plants and animals slowly, over the course of many generations, to produce changes in their forms. This "force" has no mathematical description; it is not to be found in any textbook of physics, listed alongside the basic forces which control the world of nonliving matter; but, nonetheless, it guides the course of evolution and shapes the forms of living creatures—on this planet and on all planets on which life has arisen—as firmly and as surely as gravity controls the stars and the planets.

Darwin was led to his discovery by observations of plant and animal life carried out between 1832 and 1836, during a voyage around the world on HMS *Beagle,* a British navy vessel assigned to surveying and mapping duties in the southern hemisphere. He had sailed on the *Beagle* as a naturalist, serving without pay and collecting specimens during a journey that carried him around most of the South American continent, to Australia, New Zealand, Africa and many Atlantic and Pacific islands.

Darwin suffered from seasickness throughout the five-year journey, from the day he stepped on board the *Beagle* to the day he quit her decks. He wrote from Brazil, shortly before the end of the voyage, "I loathe . . . the sea and all ships which sail upon it." And when he reached England he never boarded a ship or left his native country again. But throughout the rest of his life Darwin drew on the store of experiences accumulated in that single voyage. Toward the end of his life he wrote, "The voyage of the *Beagle* has been by far the most important event of my life. . . ."

From 1832 to 1835 the ship sailed up and down the coast of South America, and on several occasions during that time Darwin went ashore for long overland journeys of exploration. During these trips ashore Darwin

Charles Darwin: A Year Before His Death

unearthed the evidence which first turned his mind toward evolution. He came upon beds of fossils containing the skeletons of animals that had once roamed the Argentine Pampas but had been extinct for tens of thousands of years.

One of them was a toxodon, "one of the strangest animals ever discovered," the size of a rhinoceros, but with front teeth resembling those of a rodent; another was a "gigantic armadillolike animal" resembling the modern armadillo of South America, but ten times larger.

None of these extinct creatures was the same as any animal now alive on the South American continent; and yet some of them bore a surprising resemblance to existing species. Darwin thought about this "wonderful relationship . . . between the dead and the living"; was it possible that all the animals living on the earth were directly descended from vanished species? Could it be that the passage of vast amounts of time had, in some way, worked the changes between those ancient animals and their modern descendants?

Other facts impressed Darwin later in the voyage. The *Beagle* stopped for a month at the islands of the Galapagos Archipelago, situated on the equator 500 miles west of the coast of South America. During the visit Darwin noticed a "most remarkable feature" of these islands: although they were close to one another, and the climate and soil were the same on all, different kinds of plants and animals lived on them. Some of the islands even contained plants or animals which were to be found on those islands and on no others. On James Island, for example, Darwin found thirty kinds of plants which were exclusively confined to this one island, and were not to be found elsewhere in the Archipelago. He wrote, "I never dreamed that islands, about fifty or sixty miles apart . . . formed of precisely the same rocks, placed under a quite similar climate . . . would have been differently tenanted. . . ."

Darwin puzzled over the "eminently curious" nature of these variations in species; if all forms of life on the earth had been placed here by separate acts of creation, why was the creative force so prodigal in bestowing separate species on each island in the Galapagos?

Another observation led Darwin to the answer. He had also noticed that most of the animals on the islands resembled animals which were peculiar to the neighboring South American continent, and were not to be found in other parts of the world. The significance of this fact did not occur to Darwin until his return from the voyage of the *Beagle* in 1836. Then an explanation occurred to him: a long time ago plants, insects, birds, reptiles and mammals, carried by currents of air and wind, or floating on driftwood, must have reached the Archipelago from the adjacent coast. Isolated from the mainland, they evolved into forms which came to differ more and more, in the course of time, from those of their mainland cousins. Moreover, when the migrant plants and animals first arrived at the Archipelago and became established, they were identical on every island; but gradually, because the islands were isolated and cross-breeding between islands rarely occurred, distinct lines of evolution developed on each. In this way the separate islands acquired their characteristic flora and fauna.

Darwin's reasoning implied that the forms of life could change and evolve with the passage of many generations. By 1838 he was convinced that "such facts as these . . . could only be explained on the supposition that species gradually become modified." He was convinced of the truth of evolution.

These views were contrary to the opinions of many scientists and nearly all laymen, the common view being that every form of life had been specially and independently created. They were contrary also to the views held by Darwin himself when he boarded the *Beagle* in 1832. Yet six years later, Darwin found he could not ignore his own evidence on the Pampas; he had dug fossil skeletons, some closely resembling living forms, out of the ground in Argentina with his own hands; in the Galapagos he had seen with his own eyes the basic resemblance of island animal life to the life on the South American mainland; and he had seen the differences between corresponding species on separate islands in the Archipelago.

But how could Darwin convince a skeptical world that it must accept a theory which violated its basic beliefs? He wrote subsequently in his *Autobiography* regarding his belief in evolution: "The subject haunted me . . . It seemed to me almost useless to endeavor to prove by indirect evidence . . ."

So he sought a direct proof; he looked for a *cause* of evolution—a principle in nature which would make evolution a necessary and inevitable aspect of life, and, at the same time, would explain the different forms which plants and animals had assumed.

Throughout the years following the voyage of the *Beagle,* Darwin's mind turned on the problem of a cause for evolution. Gradually the outlines of a new theory emerged. It is hard to say precisely when Darwin first saw the light; no doubt the truth dawned on him slowly; but by the end of 1838 the new law of nature was clearly formulated in his notebooks. Yet he did not announce it to the world immediately; he knew that his belief in evolution would make him an unpopular figure. "It is like confessing a murder," he wrote later. In order to strengthen his case he first collected every bit of evidence which could bear on the question; he wrote in 1844, "I have read heaps of . . . books, and have never ceased collecting facts"; and in 1858, "I am like Croesus overwhelmed with my riches and facts." At last, in November, 1859, Darwin's theory appeared in print* with the title of

THE ORIGIN OF SPECIES

BY MEANS OF NATURAL SELECTION
OR
THE PRESERVATION OF FAVORED RACES IN THE STRUGGLE FOR LIFE

Darwin's apprehensions regarding the reception of his theory were confirmed immediately on its publication. The first edition of the *Origin* aroused intense interest; the entire printing sold out on the day of its appearance, and it drew down on the head of its author a storm of vituperation and ridicule such as has never greeted any other work in the history of science. After reading it, his geology professor at Cambridge wrote to him, "I laughed . . . till my sides were almost sore . . . utterly false and grievously mischievous . . . deep in the mire of folly." Other critics were less gentle; an anonymous reviewer wrote in the *Edinburgh Quarterly Review* of Darwin's "rotten fabric of guess and speculation . . . dishonorable to Natural Science."

But the argument set forth in *The Origin of Species* was beautifully simple and clear; its validity should have been apparent to everyone. Darwin began with an almost self-evident set of observations on the nature of life: All living things reproduce themselves; reproduction is the essence of life; *but the process of reproduction is never perfect.* The offspring in each generation are not exact copies of their parents; brothers and sisters differ from one another; no two individuals in the world are exactly alike, except for identical twins at the moment of birth.

Usually the variations are small; brothers and sisters resemble one another,

* Darwin might have put off publication and collected facts to the end of his life if an accident of history had not goaded him into action. For years his colleagues had urged him to publish; they had warned him that he would be anticipated if he delayed. In June, 1858, their predictions came true; Darwin received a letter from the naturalist Alfred Russel Wallace proposing a theory of evolution very clearly formulated and identical with Darwin's but arrived at independently. Friends arranged for presentation of a joint paper to the Royal Society over the names of Darwin and Wallace. Then Darwin set to work in earnest. In 1859, after "13 months and 10 days of hard labor," *The Origin of Species* appeared, and it became clear that Darwin had taken infinitely greater pains than had Wallace to collect evidence for the defense of the theory. It was the mountain of detail in the *Origin,* and Darwin's painstaking analysis of the evidence, that eventually overcame the fierce opposition to the theory and secured its acceptance.

all human beings look more or less alike, and all elephants look more or less like other elephants.

Yet, Darwin asserted, these small variations are critically important; for, in the struggle for existence, the creature which is distinguished from its brethren by a special trait, giving it an advantage in the competition for food, or in the struggle against the rigors of the climate, or in the fight against the natural enemies of its species—that creature is the one most likely to survive, to reach maturity, *and to reproduce its kind.* Some of the offspring of the favored individual will inherit the advantageous characteristic; a few will possess it to a greater degree than the parent. These individuals are even more likely to survive and to produce offspring.

Thus, through successive generations, the advantageous trait appears with ever-increasing strength in the descendants of the individual who first possessed it.

Not only does the trait become more pronounced in each individual with the passage of successive generations, but the number of individual animals possessing it also increases. For these favored individuals have slightly larger families than the average, because they and their offspring have a greater chance of survival; in each generation they leave behind a greater number of offspring than their less-favored neighbors; their descendants multiply more rapidly than the rest of the population, and in the course of many generations, their progeny replace the progeny of the animals that lack the desirable trait.

In *The Origin of Species,* Darwin gave this process the name by which it is known today: "This principle of preservation or the survival of the fittest, I have called *Natural Selection.*"

Through the action of natural selection, a favorable trait which first appeared as an accidental variation in a single individual will, with the passage of sufficient time, become a pronounced characteristic of the entire species. So the deer became fleet of foot, for the deer which ran fastest in each generation usually escaped their predators and lived to produce the greatest number of progeny for the next generation. So did man become more intelligent, for superior intelligence was of premium value: the intelligent and resourceful hunter was the one most likely to secure food. Thus developed the brain of man; thus, too, in response to other pressures and opportunities in their environments, developed the trunk of the elephant and the neck of the giraffe.

Of course, the incorporation of one new trait does not create an entirely new animal. But if we count all the births which occur to a single species over the face of the globe in one year, an enormous number of variations will appear in this multitude of young creatures. On all these variations the same process of selection works steadily, preserving for future generations the new traits which give strength to the species, and eliminating those which lend weakness. The changes may be imperceptible from one generation to the next, but over the course of many generations the accumulation of many favorable variations, each slight in itself, completely transforms the animal. According to Darwin,

> Natural selection is daily and hourly scrutinising, throughout the world, the slightest variations, rejecting those that are bad, preserving and adding up all that are good; silently and insensibly working at the improvement of each organic being in relation to its . . . conditions of life.

Natural selection molds the forms of life. Under its continuing action the shapes of animals change with time; old species disappear in response to changing conditions, and new ones arise. Few of the species of animals which roamed the face of the earth 10 million years ago still exist today, and few of those existing today will survive 10 million years hence. To quote again from *The Origin of Species*: ". . . Not one living species will transmit its unaltered likeness to a distant futurity." But natural selection works its effects subtly. Its influence is not felt in one individual or in his immediate descendants. A thousand generations may elapse before a change becomes noticeable; in man that amounts to 20,000 years. Yet, ever since Rutherford measured the age of the earth, we have known that enough time is available. Our planet has existed for billions of years; that is the secret strength of Darwin's theory. "We have almost unlimited time," he wrote in 1858, in explaining how the slightest variations in the form of an animal can grow, through their effect on the probability of producing progeny, until, after the passage of "millions on millions of generations," great changes are effected. And in the *Origin:*

> The mind cannot grasp the full meaning of the term of even a million years; it cannot add up and perceive the full effects of many slight variations, accumulated during an almost infinite number of generations. . . . We see nothing of these slow changes in progress, until the hand of time has marked the lapse of ages, and then . . . we see only that the forms of life are now different from what they formerly were.

Darwin's critics were not accustomed to thinking in terms of millions of generations and tens of millions of years; they accused him of proposing that natural selection could convert "an oyster into an orangutan" or "tadpoles into philosophers"; they taunted him with his inability to supply the missing link—the animal caught midway in the transition from one species to another. A British magazine wrote in 1861, "We defy any one, from Mr. Darwin downwards, to show us the link between the fish and the man. Let them catch a mermaid. . . ."

The hapless naturalist could not oblige them. Throughout the long battle for the acceptance of his views, Darwin was plagued continually by his inability to compress the time scale of nature and demonstrate a transformation of species to his critics. Had he known it, an example was at hand which would have provided him with the proof he needed. The case was an exceedingly rare one, in which a major evolutionary change occurred in the brief interval of fifty years.

The animal which underwent the transformation was a member of the insect world, the humble Peppered Moth, found in abundance throughout England. In the nineteenth century two varieties of this moth were known. One possessed a speckled coloration, blending perfectly into the background of lichen-covered tree trunks, which provided its normal resting place. The other variety was dark, almost black in color, and stood out conspicuously against the light background of lichens and bark. The speckled variety, known as the Peppered Moth, was the commoner form; the dark variety was readily picked off and eaten by birds, and was relatively rare.

During the course of the nineteenth century, soot progressively darkened the tree trunks of the English Midlands. As much as two tons of soot fell each day on every square mile of some industrial towns in that area. The speckled coloration of the Peppered Moth, which must originally have appeared as a chance mutation, had been developed and refined by the action of natural selection over many generations, because of its favorable effects in the struggle for survival. This same characteristic, because of a change in the environment, in this case wrought by man, now placed its possessor at a disadvantage; the Peppered Moth stood out clearly against the background of the soot-covered tree trunks, and was detected with ease by the birds of the region. The black moth, on the other hand, blended well into the new background; the trait which had formerly been unfavorable was now favorable, enhancing the chances of survival to maturity and the production of offspring. The black moth, once a rare variety, multiplied in number until it became the dominant form. The change was dramatic and swift; the first recorded capture of a dark moth took place in Manchester in 1848, and by 1900 the dark moth outnumbered the speckled variety by 99 to 1.

Even if Darwin had been able to produce the example of the Peppered Moth, it is doubtful if he would have stilled the voices of criticism, for the objections to his theory of evolution were not raised on rational grounds alone. There was also an emotional reaction to the implications of the theory for the descent of man. In the *Origin,* Darwin had deliberately avoided discussion of man's ancestry; while the book was in preparation he had written to a friend, "You ask whether I shall discuss man. I think I shall avoid the whole question. . . ." Darwin's critics were quick to supply the missing discussion: Darwin asserted that the forces of nature, acting through the struggle for survival, work continuously for the improvement of all forms of life; it followed that each animal now on the face of the earth must be descended from a related but more primitive ancestor. What animals provided a clue to man's primitive ancestry? Monkeys and apes were less advanced than man, yet closer to him in form and intelligence than any other creature; they represented primitive forms of the human being. The ridiculous monkey and the brutelike gorilla resembled man's ancestors.

Many articulate defenders of man's noble heritage entered the lists against the rash and blasphemous scientist. One of the most prominent and eloquent anti-Darwinians was Samuel Wilberforce, Bishop of Oxford. On June 29, 1860, six months after the publication of the *Origin,* 700 people crowded into a hall in Oxford University to hear Bishop Wilberforce debate the merits of Darwin's theory with the biologist, Thomas Huxley, who had become Darwin's most ardent supporter. Toward the end of the debate, Bishop Wilberforce turned to Huxley and asked, "Was it through his grandmother or his grandfather that he claimed his descent from a monkey?"

Huxley's reply is one of the most famous ripostes in the history of science. Whispering to his neighbor, "The Lord hath delivered him unto my hands," he rose and said, "If I am asked whether I would choose to be descended from the poor animal, of low intelligence and stooping gait, who grins and chatters as we pass—or from a man, endowed with great ability and a splendid position, who would use these gifts to discredit and crush humble seekers after truth, I hesitate what answer to make."

The arguments over Darwin's views gradually subsided through the decade of the 1860s. They erupted again when the *Descent of Man* was published in 1871. In this book, Darwin presented his views on man's origin and history, confirming the darkest suspicions of his critics by setting forth evidence linking man and the apes to a common ancestor. But by Darwin's death in 1882, his theories were widely accepted in the scientific world, and had made a substantial impact on the thinking of all men. Today the basic concepts of the Darwinian theory of evolution have few opponents.

AN EXAMPLE OF NATURAL SELECTION. In *The Origin of Species* Darwin asserted that the forms of living creatures are shaped by the struggle for survival. The individuals born with special traits, enabling them to compete for food, resist the rigors of the climate, and escape from predators, are the ones most likely to survive to maturity and to reproduce. Their favorable traits are inherited and spread throughout the population in the course of many generations.

Usually the pace of these evolutionary changes is too slow to be observed within the span of one lifetime, but in the case of the Peppered Moth of England, a major evolutionary change occurred in fifty years. In its original form this moth possessed speckled gray-and-white markings, which blended perfectly into the lichen-covered trunks of trees, protecting it from detection by birds. A speckled specimen appears in photograph 1A, barely visible against the tree trunk in the lower right-hand side of the picture. In the same photograph appears a second variety, very dark in color because its body chemistry manufactured an excessive amount of melanin. In older times the dark specimens were rare because they stood out easily from the barks of trees and were readily detected and eaten by birds (1B).

In the aftermath of the Industrial Revolution, tree trunks throughout large areas of England were blackened by soot, against which the speckled Peppered Moth stood out conspicuously (2A). The dark variety, which appears at the bottom of 2A, is nearly invisible against the soot-blackened tree. This change in environment led to the rapid decimation of the Peppered Moth (2B); the dark variety rapidly became the dominant species, outnumbering the original form of this moth by 100 to 1 in a census taken in 1900.

Many species of moths and butterflies in England are becoming darker as a consequence of changes in their environment wrought by industrialization—changes that have converted a formerly unfavorable mutation, the excess of melanin, to a favorable mutation in the course of only fifty years.

1A

1B

2A

2B

14 DNA and Darwin

THROUGHOUT the years in which Darwin's ideas were winning increasing acceptance, one point remained obscure. What was the origin of the variations from one individual to another, which provided the raw material for natural selection? Regarding these variations, which played so essential a role in his theory, Darwin could only say helplessly, "We are profoundly ignorant of the cause of each slight variation or individual difference. . . .[They] seem to us in our ignorance to arise spontaneously." The ignorance was not fully dispelled until 1953, nearly a century after the publication of the *Origin,* when it became clear how the basic characteristics of the individual are passed from generation to generation. These characteristics reside in the molecule called DNA, which is found in the cells of every living organism on the earth. The DNA molecule is a long chain of the smaller molecules called nucleotides, which are arranged in a sequence special to each organism. As we have noted, no two individuals in the world, except identical twins, have the same sequence of nucleotides in their DNA. This sequence determines which proteins will be assembled in the cells of the body; and the proteins, in turn, control the body chemistry and all the traits of the individual. Thus, the DNA in every creature contains the master plan for that creature.

We now know that occasionally some of the nucleotides in the DNA molecule are damaged, altered or removed entirely from the molecule, so that the master plan is changed. The damage or alteration may affect only one nucleotide in the long chain—a chain which may, in human cells, stretch out over a billion nucleotides. Nonetheless, the change in a single nucleotide may be critically important, for the proteins in the cell are assembled out of amino acids in a sequence that follows the order of the nucleotides in the DNA. Damage to one of these nucleotides, or replacement of one nucleotide by another of a different type, will lead to the assembly of a different protein, in which, at one point along the chain of amino acids making up the protein, the wrong kind of amino acid is located.

Sometimes the modified protein is able to play its normal role in the chemistry of the cell. At other times, when the improperly placed amino acid is located at a critical site, the effectiveness of the protein is destroyed.

When a modified DNA and the modified protein produced by it are situated in an ordinary cell of the body, the abnormal cell is soon replaced by the growth of new cells and the effect of the change in the DNA molecule quickly disappears. However, in one type of cell in the body a change in the sequence of nucleotides in the DNA molecule may have permanent and serious consequences. This is the germ cell—sperm in the male and ovum in the female. Like all other cells, the germ cell contains its set of DNA molecules with the master plan for the development of the individual. When the sperm and ovum unite to form a fertilized egg, every organ in the body of the mature individual subsequently develops out of the egg by repeated cell division, following a new, joint master plan provided by the combination of the DNA molecules from the sperm and the ovum participating in the union. If the DNA in one of these two germ cells has been damaged or altered, the effect of the change will appear in every cell of the body of the new individual. Moreover, it will be transmitted to the offspring of this individual in the next generation, and in every generation thereafter. All the descendants of that individual, down through the corridors of time, will bear the mark of the change in the sequence of nucleotides in the ancestral DNA.

A modification of the nucleotides in the DNA of a germ cell is called a *mutation.* Mutations are changes in the body chemistry of the individual *which are transmitted to his progeny.* They are the inheritable variations which form the basis of Darwin's theory of evolution.

Some mutations change the chemistry of the body in a way which improves the chances of the individual for survival; these are called *favorable* mutations. The individuals possessing a favorable mutation are always the ones most likely to propagate their species; from generation to generation their number steadily increases, and in the course of many generations the favorable mutation spreads throughout the population. Mutations may also be *unfavorable,* diminishing the chance of survival to maturity and, therefore, the chance of producing offspring. These mutations are gradually weeded out of the population. That is the way in which evolution works: By the pruning away of unfavorable mutations simultaneously with the strengthening of favorable ones.

What is the cause of mutations? What can damage or modify the sequence of nucleotides in the germ cells of an organism? When we know the answer to this question, we are close to an understanding of the cause of evolution.

One cause of mutations lies in the organism itself. Occasionally an error—an imperfection—occurs in the copying process by which the DNA molecule reproduces itself. At the start of the process of reproduction, just prior to the division of a cell into two daughter cells, the double-stranded DNA in the parent cell unwinds and separates into two single strands. Each strand gathers new nucleotides from the pool of nucleotides floating in the cell to form a new, double-stranded DNA duplicating the original. It is at this point that the copy error may occur. One of the newly added nucleotides may be of the wrong kind; that is, it fails to match its counterpart in the existing strand. As a result, when the assembly of the two DNA molecules has been completed, the sequence of nucleotides in one of the daughter DNAs differs from the sequence of nucleotides in the parent DNA. That daughter DNA has suffered a mutation.

Copy errors are not the only source of mutations. The DNA molecule can also be altered by chemicals, if they enter the blood stream. Mustard gas, the poison gas of the First World War, is effective in this way. LSD appears to cause serious damage to the DNA molecule. And DNA can be altered by particles or radiation that are energetic enough to penetrate the body. The physician's X-ray machine is one source of penetrating radiation. Nuclear bomb explosions are another. In addition to these man-made sources, there are also cosmic rays —particles produced by unknown forces in distant regions of the universe— which bombard the earth from all directions. Occasionally, either a cosmic ray or radiation produced by a man-made source will pass through a germ cell, even if the cell is buried deeply in the body, and will disrupt the normal sequence of nucleotides in its DNA, producing a mutation.

Which of these sources is the prime cause of mutations? Mustard gas, X-ray machines and nuclear bombs are recent products of man's ingenuity; they have not been with us long enough to have had an appreciable influence on the course of evolution. But copy errors have occurred since the beginning of life on the earth; and cosmic rays probably have existed since the beginning of time. We can guess that these latter two sources have played roles of comparable importance in the past history of the evolutionary process. Whether the rate of mutation and the future rate of evolutionary change will be increased appreciably by the other, man-made sources of mutation, and whether the augmentation will improve or weaken the human species—these are open questions.

☼

15 The Ascent of Man

IN THE course of 3 billion years, life on the earth evolved from a soup of organic molecules to the carnival of animals that now plays across the face of the planet. Among these animals is man. By what sequence of events did he arise out of a broth of DNA and proteins? What circumstances guided the course of evolution from the first primitive organisms to the highest expression of life in the form of the human being?

The history of these events probably began with the appearance of the first self-copying molecules in the waters of the earth. These molecules were similar to DNA, and may have been identical with it. The waters also contained amino acids. In some way, which we have not yet been able to reconstruct in the laboratory, those self-copying molecules developed the ability to serve as guides for the assembly of amino acids into proteins. At first the DNA molecules were short strands containing only a few nucleotides, and could assemble only simple proteins. In the course of time they evolved into longer chains, capable of assembling complex proteins of many kinds. Some of these proteins were primitive enzymes; they hastened the chemical reactions which led to the growth and reproduction of DNA. Other proteins were structural; they were of the kind out of which cell walls were formed.

With the appearance of the structural proteins, a new advance became possible in the organization of living matter. The DNA molecule came to reside in the center of a cell, whose wall was a porous membrane which permitted small molecules, such as amino acids and nucleotides, to pass through from the surrounding fluid to the interior, but did not permit the larger molecules, such as DNA and the proteins assembled under its control, to pass out again in the reverse direction.

The primitive development of the cell, concentrating in the vicinity of the DNA all the chemicals needed for growth and reproduction, marked the greatest single step ever taken in the course of evolution. Several hundred million or, possibly, a billion years must have been required for the evolution of the cell; but once it appeared, this efficient form of life must have spread rapidly throughout the waters of the earth, engulfing and replacing all the cell-less molecules which preceded it.

We can assume that in a relatively short time—perhaps within 100 million years—the one-celled organism evolved into a colony of cells. With the further passage of time, groups of cells within those colonies assumed specialized functions of food-gathering, digestion, the structural features of an outer skin, and so on; thus began the stage of evolution leading to the complex, many-celled creatures which dominate life today.

The fossil record contains no trace of these preliminary stages in the development of many-celled organisms. The first clues to the existence of relatively advanced forms of life consist of a few barely discernible tracks, presumably made in the primeval slime by soft, wriggling wormlike animals. These are found in rocks about one billion years old. Somewhat later, well-defined worm burrows appear in the record. These meager remains are the earliest traces of many-celled animal life on the planet.

Little else appears in the fossil record during those first several billion years. One of the mysteries in the study of life is the fact that suddenly, in rocks 600 million years old, the record explodes in a profusion of living forms. A great variety of animals appears in the record at that time. Perhaps the forms of life were nearly as numerous and populous just prior to this magical date, but left no trace of their existence because they lacked the body armor which is most easily preserved.

Somewhat more than 400 million years ago an event occurred which is of great consequence for the development of man. There appeared, for the first time, a new kind of creature—one with an internal skeleton and a backbone. This animal—the vertebrate—evolved out of a wormlike ancestor resembling the modern lancelet, a small, translucent creature, lacking fins and jaws, but possessing gills and, most important, a primitive version of the backbone.

Among the descendants of the first vertebrates were the fishes. Some of the early fishes contained crude lungs for gulping air at the surface of the water, as well as gills. These lungs were lost or converted to other uses in most instances, but in some forms of fish, perhaps those living in small bodies of water such as ponds and tidepools, the lungs came into frequent use. Whenever a drought developed and the water level in the ponds dropped, the fish with the best lung capacity survived where others perished. They lived to produce progeny which inherited their superior capacity for breathing air. In this way, an efficient lung evolved gradually among the fish inhabiting shallow bodies of water.

Some of the air-breathing fish were doubly favored in possessing strong fins which enabled them to waddle over the land from one pond to another in search of water. By a slow accumulation of favorable mutations, the muscle and bone of the fin gradually changed into a form suitable for walking on land. In this way, the fin evolved into the leg. The metamorphosis took place over a period of perhaps 50 million years, and a like number of generations. The result was a four-legged, air-breathing animal, known as the amphibian.

The amphibian was still tied to the water, because its skin required frequent moistening; also, its eggs, like those of fish, lacked a hard casing. If deposited on land they dried out and the embryo died. Therefore the eggs of the amphibian had to be laid in water or in moist places.

The amphibians were born in water, lived most of their adult lives near water, and almost always returned to the water to lay their eggs. For fifty million years they flourished on shores and river banks. Some became large, aggressive carnivores as much as 10 feet in length, fearing no other animals of their time. The amphibians attained the peak of their size 250 million years ago, and thereafter they declined. Today their common descendants are the diminutive frog, toad and salamander.

In the course of time, some of the ancient amphibians, again by the chance occurrence of a succession of favorable mutations, developed the ability to lay their eggs on land. These eggs were encased in a firm, leathery shell, which retained moisture and provided the embryo with its own private pool of fluid. Other mutations led to a tough, leathery hide, which preserved the water in their bodies without the need for continual immersion. Such creatures were completely emancipated from the water. They were the first reptiles.

The reptiles marked a very succesful step in evolution, for they had access to rich resources of food previously denied to the fishes and the amphibians. The reptiles flourished and developed into a great variety of forms, including the ancestors of every land animal with a backbone now on this earth. They reached their evolutionary zenith in the dinosaurs. These animals ruled the earth for 100 million years. They displayed an extraordinary vigor, evolving into such extreme forms as the giant vegetarian swamp-dweller, *Brontosaurus,* 70 feet long and weighing 30 tons; and the meat-eating *Tyrannosaurus rex,* 40 feet high, with a 4-foot skull filled with daggerlike teeth—unquestionably the fiercest land-living predator the world had ever seen.

Two hundred million years ago, somewhat before the appearance of the first dinosaurs, another branch of the reptile class veered off on an entirely different course. This particular group may have lived in places on the edge of the temperate zone where the weather was relatively severe. Through the action of natural selection on chance variations, the new branch of the reptiles acquired a set of traits which fitted them uniquely for survival in a rigorous climate. They developed the rudimentary characteristics of a warm-blooded animal. The naked scaly skin of the reptile was replaced in these animals by insulating coats of hair and fur which kept them warm in low temperatures, while sweat glands under the skin, controlled by an internal thermostat, cooled the body by evaporation when the temperature rose too high.

These traits developed slowly over the course of tens of millions of years. Other changes took place at the same time. Many four-footed reptiles had a clumsy, sprawling posture with the legs spread out from the trunk; rapid movement was impossible with a skeleton so constructed. In the new line of evolution the legs pulled in beneath the body, raising it from the ground and permitting a fast, running gait. Important modifications appeared in the teeth: near the front of the mouth were two large canines suitable for tearing off pieces of the prey; behind these were cusped teeth resembling molars, for cutting and grinding the food down to a smaller size. A quick replenishment of energy and a high level of activity are possible with such teeth, in contrast to the postprandial stupor of the reptile that has swallowed its prey whole.

These animals, resembling a cross between a lizard and a dog, constituted the ancestral stock of the mammals.

Modern mammals possess other characteristics, in addition to warm-bloodedness, which distinguish them from their reptilian forebears; the most important among these is an exceptionally effective means, unique to mammals, of caring for their young. Reptiles lay their eggs and commonly display no further interest in the fortune of their progeny. Birds are better parents; they care for their eggs, but the defenseless, unhatched embryo is, nonetheless, often the victim of egg-hungry creatures; and, moreover, the newly hatched fledgling must be left to the mercies of predators while its parents search for food. The mammalian mother nourishes the developing embryos of the unborn young inside her body, where they are well protected against hostile elements in the environment; after birth, she nourishes her young with her milk, secreted by the glands which have given the mammals their name; and she continues to care for the young a long time thereafter, until they are able to fend for themselves. The mammals make more effective provisions for the survival of their young than any other animal, thereby securing a very great advantage in the competition for the propagation of their species.

In spite of these special talents, the mammals remained subordinate to the dinosaurs for more than 100 million years—small, furry animals, inconspicuous, keeping out of sight of the rapacious reptiles by living in the trees or in the grasses.

But seventy million years ago, the dinosaurs died out. The reasons for their disappearance are still obscure. It is likely that their downfall was the consequence of a worldwide change in climate, which they were ill-equipped to survive. Dinosaurs, like all reptiles, were cold-blooded animals; that is, they lacked the internal heat controls which could maintain the temperature of the body at a constant level regardless of the rigors of the climate. We know that the period during which they disappeared was marked by repeated upheavals of the earth's crust, in which many new mountain ranges were formed. The Rocky Mountains were among the ranges created in these upheavals. Most probably, the upward thrust of huge masses of rock disrupted the flow of currents of air around the globe; perhaps the climate of the temperate zone was changed in this manner from one of uniform warmth and humidity, agreeable to a cold-blooded animal, to a climate marked by major changes of temperature from season to season.

As the population of dinosaurs dwindled, the mammals came down from the trees and up from their burrows in the ground, and they inherited the earth. Quickly they spread out across all the continents. Within 20 million years, the basic mammalian stock evolved into the forebears of most of the mammals with which we are familiar today—bats, elephants, horses, whales and many others.

But one group of mammals remained in the trees. These mammals—the primates—were singled out, by the circumstance of their tree-dwelling existence, to be the ancestors of man. They were small, insect-eating animals, the size of a squirrel, and similar in appearance to the modern tree shrew of Borneo. Man owes his remarkable brain to the fact that these animals required two physical attributes for survival in their arboreal habitat: First, they needed hands and an opposable thumb for securing a tight hold on branches; and second, they needed sharp binocular vision to judge the distances to nearby branches. In the competition for survival among primitive tree-dwelling mammals, 100 million years ago, those who possessed these characteristics in the highest degree were favored. They were the individuals most likely to survive and to produce offspring. Through successive generations the desirable traits of a well-developed hand and keen vision, passed on from parents to offspring, were steadily refined and strengthened. By 50 million B.C. they already appeared in advanced form in the animals from which the modern tree shrew, lemur and tarsier are descended. They became even better developed in some of the immediate descendants of these animals, under the continued pressure of the struggle for survival in the trees. Gradually, the evolutionary trends established by the requirements of life in the trees transformed some of these early primates into animals resembling the monkey.

Animals with hands also had the potential capacity to exercise rudimentary manual skills; when this potential was combined with the development of the associated brain centers, such animals had, almost by accident, the ability to use tools. For those who had this ability, great value became attached to the mental capacity for the remembrance of the usage of tools in the past, and for the planning of their use in the future; thus, by the action of natural selection on a succession of chance mutations, those centers of the brain developed and expanded in which past experiences were stored and future actions were contemplated. These mental qualities proved to be of great value in meeting the general problems of survival. As a result, the brain evolved and expanded under the continued pressure of the struggle for existence. It doubled in size in 10 million years, and nearly doubled again in the next million. Thus was the line of ascent leading to man firmly established.

By this chain of evidence and theory, the distinguishing characteristic of the human condition—intelligence—may be traced back to the accidental circumstance of a tree-dwelling ancestry.

The path of evolution stretches further back into time—from the tree-dwelling forebears of man to the first mammal; then to a doglike reptile of a kind that no longer exists; to the first vertebrate; from the vertebrates to a succession of soft-bodied animals lost in the sands of history; then across the threshold of life into the world of nonliving matter; and finally, many billions of years ago, long before the solar system existed, into the parent cloud of hydrogen.

There is grandeur in this view of life, with its several powers,
having been originally breathed by the Creator into a few forms
or into one; and that, whilst this planet has gone cycling on
according to the fixed laws of gravity, from so simple a beginning
endless forms most beautiful and most wonderful have been
and are being evolved.

—CHARLES DARWIN
The Origin of Species

EARLY FORMS OF LIFE. Among the oldest residues of living organisms are the stromatolites, deposits left by primitive algae approximately 3 billion years ago. The photograph above shows an algal stromatolite found in the Precambrian rocks of the Medicine Bow Mountains in Wyoming. On the opposite page are other remains of early life. The residues and simple fossils on these pages are typical of the earliest traces of life discovered thus far.

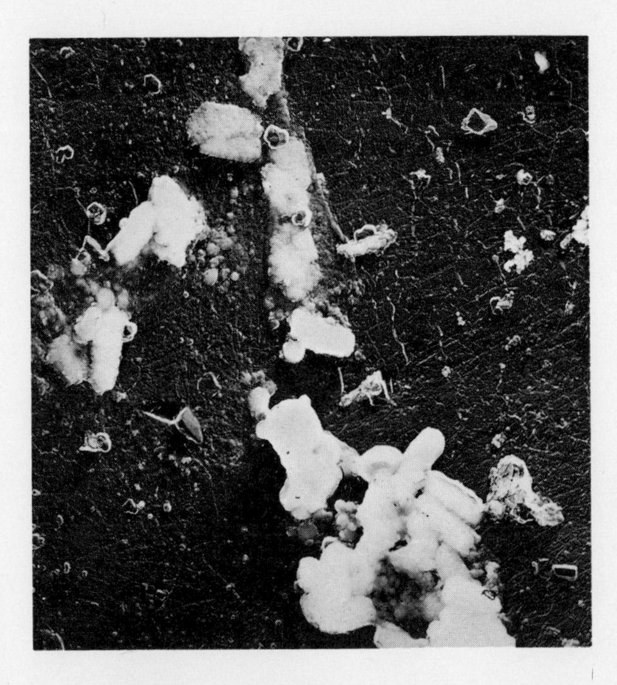

The fossilized remains of bacteria approximately 2 billion years old *(left)*.

Burrows made by worms that lived approximately 1 billion years ago *(below)*.

THE EVOLUTION OF THE VERTEBRATES. Four hundred million years ago the first animal with a backbone—a primitive fish—appeared on the earth. Some of the early fishes developed the ability to breathe air and to crawl up onto the land. From these creatures developed the amphibians and then the reptiles. Reptiles, the first backboned animals to be completely emancipated from the sea, spread out over the land and flourished there for 150 million years. Some of the reptiles were the ancestors of the modern snake, lizard and turtle; others evolved into the dinosaurs. One line led to the ancestors of the modern birds and, about 200 million years ago, other branches evolved into the mammals. Both birds and mammals were distinguished from the reptiles in one critical way: they were warm-blooded. This characteristic enabled them to withstand extremes of hot and cold in the external environment. Sixty million years ago, when the dinosaurs suddenly disappeared, probably because of severe changes in climate, the mammals inherited the earth and evolved into myriad forms. One branch of the mammals—the primates—were the ancestors of man.

A RELATIVE OF MAN. The tarsier *(above)* resembles the small tree-dwelling mammals that were the progenitors of man. In order to survive in the trees these animals needed grasping hands and binocular vision. The possession of these characteristics stimulated the development of the brain centers that coordinated sight and touch. In the struggle for existence 50 million years ago those creatures were selected for survival in whom these traits were the most highly developed. Gradually the tarsier-like animals evolved into the ancestor of the monkey, and then into an ape-like creature which was the forebear of man. Manual dexterity and the development of the associated brain centers, traits required for an arboreal existence, gave their possessors, almost by accident, the potential ability to use tools and to remember past usage of tools. Thus developed the parts of the brain in which past experiences were stored and future plans made. Man owes his distinguishing feature—his intelligence—to the circumstance that his ancestors lived in trees.

Picture Credits

Frontis.	Photograph from the Mount Wilson and Palomar Observatories
Page	
4	Barrett Gallagher
10	The Bettmann Archive
12	Photograph from *Rutherford and the Nature of the Atom,* by E. N. da C. Andrade. Copyright 1964 by Educational Services Incorporated. Published by Doubleday & Company, Inc.
14	Adapted from *The Atom and the Earth,* by J. G. Navarra and J. E. Garone, by permission of Harper & Row
15, *top*	Adapted from *The Atom and the Earth,* by J. G. Navarra and J. E. Garone, by permission of Harper & Row
15, *bottom*	Adapted from *Introduction to Physical Geology,* by Chester R. Longwell and Richard F. Flint, by permission of John Wiley & Sons
16	Adapted from the Reader's Digest *Great World Atlas,* by permission of the Reader's Digest Association
18 *and* 19	Adapted from the Reader's Digest *Great World Atlas,* by permission of the Reader's Digest Association
20	Adapted from *The Universe,* by Otto Struve, by permission of the MIT Press, Cambridge, Mass.
21	Yerkes Observatory Photograph
22 *and* 23	Lund Observatory, Lund, Sweden
24	Max Gschwind for *Fortune*
25	Photograph from the Mount Wilson and Palomar Observatories
27	Photograph from the Mount Wilson and Palomar Observatories
33	Adapted from *Today's Basic Science,* Book No. 6, by J. G. Navarra and J. E. Zafforoni
37	From *Rutherford and the Nature of the Atom,* by E. N. da C. Andrade. Copyright 1964 by Educational Services Incorporated. Published by Doubleday & Company, Inc.
48	Adapted from *Atlas of the Universe,* by Nelson and Elsevier
49	Photograph from the Mount Wilson and Palomar Observatories
51	Photograph from the Mount Wilson and Palomar Observatories
52 *and* 53	Barrett Gallagher
58	Barrett Gallagher
59	Adapted from *Atlas of the Universe,* by Nelson and Elsevier
60 *and* 61	Barrett Gallagher
70 *and* 71	Barrett Gallagher
82, *top*	Photograph from the Mount Wilson and Palomar Observatories
82, *bottom*	Department of the Air Force
83	Photograph from the Mount Wilson and Palomar Observatories
84	National Aeronautics and Space Administration
85	National Aeronautics and Space Administration
86	National Aeronautics and Space Administration

Page	
87	National Aeronautics and Space Administration
88	National Aeronautics and Space Administration
89	National Aeronautics and Space Administration
90 *and* 91	National Aeronautics and Space Administration
92	National Aeronautics and Space Administration
93	National Aeronautics and Space Administration
94 *and* 95	National Aeronautics and Space Administration
96	National Aeronautics and Space Administration
97	National Aeronautics and Space Administration
111	Lowell Observatory
112	National Aeronautics and Space Administration
113	National Aeronautics and Space Administration
114	E. M. Antoniadi, *La Planète Mars*
115	Lowell Observatory
116 *and* 117	National Aeronautics and Space Administration
118 *and* 119	National Aeronautics and Space Administration
121	Photograph from the Mount Wilson and Palomar Observatories
126 *and* 127	Spence Air Photo
128	Courtesy of the Century Co.
129	The Bettmann Archive
137, *top*	Wide World Photos
137, *bottom*	Leo Stashin for *Fortune*
143	Richard Fowler
144 *and* 145	Barrett Gallagher
148	Adapted from *Early Evolution of Life,* by Richard S. Young and Cyril Ponnamperuma, by permission of D. C. Heath and Company. Copyright © 1964 by American Institute of Biological Sciences
149	United Press International
150, *top*	C. A. Knight and R. C. Williams, Virus Laboratories, University of California
150, *bottom*	Carl T. Mattern and Herman Dubuy, National Institutes of Health, Bethesda, Maryland
151	Dr. Giuseppi Penso, Istituto Superiore de Sanita, Rome
158	Courtesy of the Century Co.
166 *and* 167	From the experiments of Dr. H. B. D. Kettlewell, Oxford University
178	*The Fossil Book,* by Carroll Lane Fenton and Mildred Adams Fenton, Doubleday & Co., Inc.
179, *top*	J. William Schopf, Elso S. Barghoorn, Morton D. Maser, and Robert O. Gordon
179, *bottom*	*The Fossil Book,* by Carroll Lane Fenton and Mildred Adams Fenton, Doubleday & Co., Inc.
180	Adapted from *The Age of Reptiles,* by Edwin H. Colbert, by permission of W. W. Norton & Company, Inc. Copyright © 1965 by Edwin H. Colbert
181	By permission of the Trustees of the British Museum (Natural History)

Index

(Numbers in italics refer to pages on which important items are introduced or defined.)

About the Author

Robert Jastrow was born in New York City in 1925. He received his B.A. from Columbia University in 1944, his M.A. in 1945, and his Ph.D. in theoretical physics in 1948. He was a postdoctoral fellow at Leiden University (1949–1950), worked at the Institute for Advanced Study in Princeton (1949–1950), and has taught at Yale and Columbia. From 1954 to 1958 Dr. Jastrow was a consultant in nuclear physics at the United States Naval Research Laboratory in Washington, D.C. He joined NASA at the time of its formation in 1958 and established the Theoretical Division of the Goddard Space Flight Center. He was chairman of the Lunar Exploration Committee of NASA from 1959 to 1961. Dr. Jastrow is now the Director of the Goddard Institute for Space Studies of NASA and Adjunct Professor of Geophysics at Columbia University.

Dr. Jastrow received the Columbia University Medal for Excellence in 1962, the Arthur S. Flemming Award for outstanding service in the U.S. Government in 1964, and the NASA Medal for Exceptional Scientific Achievement in 1968. He is editor of the *Journal of the Atmospheric Sciences,* a member of the International Academy of Astronautics, and a Fellow of the American Association for the Advancement of Science.